C++之旅

（第3版）（英文版）

[美] 本贾尼·斯特劳斯特鲁普　著

電子工業出版社
Publishing House of Electronics Industry
北京•BEIJING

内 容 简 介

本书一共 19 章，以 C++20 为标准，讲述了现代 C++所提供的编程特性。

有其他语言编程经验的读者可以从本书中快速了解 C++所具备的功能，从而获得对现代 C++更全面的认知，以便更好地了解现代 C++语言已经发展到的程度。资深程序员可以从本书作者的整体行文风格中感受到他在设计 C++特性时的一些考量及侧重点，了解 C++语言在历史上曾经历过的变迁，以及一部分特性为什么会是今天这个样子。

所以，本书适合的读者：有其他语言编程经验，想要了解 C++语言的读者；有传统 C++编程经验，想要了解现代 C++语言特性的读者；有较丰富编程经验且想了解“C++之父”在 C++设计过程中的一些设计细节与思路的读者。

图书在版编目（CIP）数据

C++之旅 = A Tour of C++：第 3 版：英文 /（美）本贾尼 • 斯特劳斯特鲁普著. —北京：电子工业出版社，2024.2

ISBN 978-7-121-47250-3

Ⅰ. ①C… Ⅱ. ①本… Ⅲ. ①C 语言－程序设计－英文 Ⅳ. ①TP312.8

中国国家版本馆 CIP 数据核字（2024）第 034111 号

责任编辑：符隆美
印　　刷：北京雁林吉兆印刷有限公司
装　　订：北京雁林吉兆印刷有限公司
出版发行：电子工业出版社
　　　　　北京市海淀区万寿路 173 信箱　邮编：100036
开　　本：787×980　1/16　印张：18.5　字数：343.36 千字
版　　次：2024 年 2 月第 1 版（原书第 3 版）
印　　次：2024 年 2 月第 1 次印刷
定　　价：109.00 元

凡所购买电子工业出版社图书有缺损问题，请向购买书店调换。若书店售缺，请与本社发行部联系，联系及邮购电话：（010）88254888，88258888。

质量投诉请发邮件至 zlts@phei.com.cn，盗版侵权举报请发邮件至 dbqq@phei.com.cn。

本书咨询联系方式：faq@phei.com.cn。

前言

在你进行指示时，简短些！

——西塞罗[1]

现代C++给人感觉像一种新的语言。我是说，相比C++98或C++11的时代，现在我能够更清晰、更简单、更直接地表达我的想法。不但如此，现代C++生成的程序也更容易被编译器检查，而且运行得更快。

本书展示了C++20定义的C++的概况，它是当前ISO C++的标准，并且已被主流C++提供商实现。另外，本书还提到了一些目前已使用的库组件，但它们还没被纳入C++23标准的计划。

就像其他的现代编程语言一样，C++也很“大”，因为它需要大量的库来提高自身的效率。这本薄薄的书旨在让有经验的程序员了解现代C++是由什么构成的，它涵盖了主要的语言特性和主要的标准库组件。本书可以在一两天内读完，但要写出好的C++代码，显然需要比读本书多得多的学习时间。然而本书的目标不是让你精通语言，而是提供概述与关键示例来帮助你着手学习。

你最好已经有一些编程经验。如果不是这样，请考虑先阅读相关的资料再继续阅读本书，推荐的资料有《C++程序设计原理与实践》（第2版）[Stroustrup, 2014]。即使你以前编写过程序，使用的语言及写过的程序与这里介绍的C++风格也可能存在非常大的区别。

想象一下，在哥本哈根或者纽约等城市观光旅游。在短短几小时内，你快速地浏览了当地主要景点，聆听了一些背景故事，并获得了一些下一步该做什么的建议。但你并不能在这样简短的旅程中完全理解这座城市，也不能完全理解所见所闻，有些故事听起来可能很奇特，

---

① 古罗马著名哲学家、学者。

甚至不可思议。你也不知道如何驾驭管理城市生活的规则，不管是正式的还是非正式的。要想真正理解一座城市，你必须在这个城市住上几年。然而，如果足够幸运的话，你可能会了解一些概况，对这个城市的特殊之处形成概念，并且对其中的一部分产生兴趣。在这次旅行结束之时，真正的探索才刚刚开始。

《C++之旅》介绍了 C++语言的主要特性，它们都支持面向对象和泛型编程之类的编程风格。不要指望本书会像参考手册那样，逐个特性地详细介绍语言的全貌。在这本最经典的教科书中，我试图在使用一个特性之前对它做出解释的，但其实很难完全做到这样，因为并不是每个人都严格按章节顺序阅读。我认为本书的读者在技术上已经非常成熟。因此，读者不妨对交叉引用善加利用。

同样，《C++之旅》对标准库的介绍以示例的形式点到为止，不会详尽描述所有细节。读者应根据需要搜索额外的资料来获取技术支持。C++生态系统涵盖的范围远超 ISO 标准提供的配套工具（例如，库、构建系统、分析工具和开发环境），读者可在网上获得海量（但良莠不齐）的资料。大多数读者可以从 CppCon 和 Meeting C++等会议中发现有用的教程和简要介绍的视频。如果读者想要了解有关语言的技术细节和 ISO C++标准提供的库，我推荐 Cppreference 网站。例如，当遇到一个标准库函数或类时，很容易就能在该网站查到它的定义，而且通过查阅它的文档，可以找到许多相关联的工具。

《C++之旅》呈现出来的 C++是一个集成的整体，而不是整齐地堆叠在一起的层状蛋糕。因此，具体的语言特性究竟是来自 C、C++98，还是来自更高版本的 ISO 标准，我极少做出标注。此类信息可在第 19 章中找到。我专注于基础知识并尽量保证内容简明扼要，但我并没有完全抵制住过度呈现新特性的诱惑，模块（3.2.2 节）、概念（8.2 节）和协程（18.6 节）这三节就是“例证”。我对最新进展的稍许偏爱，似乎也正好满足许多已经了解某些旧版本 C++知识的读者的好奇心。

编程语言参考手册或标准只是简单地说明了可以做什么，但程序员通常更感兴趣的是学习如何更好地使用该语言。鉴于此，本书所涵盖的主题是精心挑选的，在文字内容上也有所体现，尤其是在建议性章节。关于现代 C++如此优秀的原因，可以在 *C++ Core Guidelines* [Stroustrup, 2015]中找到更多观点。如果想进一步探索本书提出的理念，可以将 *C++ Core*

Guidelines 视为一个很好的参考来源。你可能会注意到，*C++ Core Guidelines* 和本书在建议的提法及建议的编号上有着惊人的相似之处，其原因之一是，《C++之旅》的第 1 版正是 *C++ Core Guidelines* 初版内容的主要来源。

鸣谢

感谢所有帮助完成和更正《C++之旅》早期版本的人，特别是在哥伦比亚大学参加我的“Design Using C++”课程的学生。感谢摩根士丹利给我时间编写本书。感谢 Chuck Allison、Guy Davidson、Stephen Dewhurst、Kate Gregory、Danny Kalev、Gor Nishanov 和 J.C. van Winkel 审阅本书，并提出了许多改进建议。

我使用 troff 完成本书的排版，并使用了 Brain Kernighan 原创的宏。

本贾尼·斯特劳斯特鲁普

于纽约曼哈顿

目录

1

The Basics
（基础）

The first thing we do, let's
kill all the language lawyers.
– Henry VI, Part II

1.1 Introduction（引言）

This chapter informally presents the notation of C++, C++'s model of memory and computation, and the basic mechanisms for organizing code into a program. These are the language facilities supporting the styles most often seen in C and sometimes called *procedural programming*.

1.2 Programs（程序）

C++ is a compiled language. For a program to run, its source text has to be processed by a compiler, producing object files, which are combined by a linker yielding an executable program. A C++ program typically consists of many source code files (usually simply called *source files*).

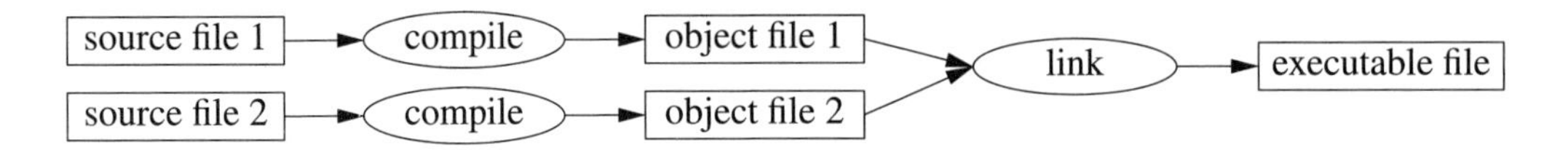

An executable program is created for a specific hardware/system combination; it is not portable, say, from an Android device to a Windows PC. When we talk about portability of C++ programs, we usually mean portability of source code; that is, the source code can be successfully compiled and run on a variety of systems.

The ISO C++ standard defines two kinds of entities:

- *Core language features*, such as built-in types (e.g., **char** and **int**) and loops (e.g., **for**-statements and **while**-statements)
- *Standard-library components*, such as containers (e.g., **vector** and **map**) and I/O operations (e.g., **<<** and **getline()**)

The standard-library components are perfectly ordinary C++ code provided by every C++ implementation. That is, the C++ standard library can be implemented in C++ itself and is (with very minor uses of machine code for things such as **thread** context switching). This implies that C++ is sufficiently expressive and efficient for the most demanding systems programming tasks.

C++ is a statically typed language. That is, the type of every entity (e.g., object, value, name, and expression) must be known to the compiler at its point of use. The type of an object determines the set of operations applicable to it and its layout in memory.

1.2.1 Hello, World!

The minimal C++ program is

```
int main() { }          // the minimal C++ program
```

This defines a function called **main**, which takes no arguments and does nothing.

Curly braces, **{ }**, express grouping in C++. Here, they indicate the start and end of the function body. The double slash, **//**, begins a comment that extends to the end of the line. A comment is for the human reader; the compiler ignores comments.

Every C++ program must have exactly one global function named **main()**. The program starts by executing that function. The **int** integer value returned by **main()**, if any, is the program's return value to "the system." If no value is returned, the system will receive a value indicating successful completion. A nonzero value from **main()** indicates failure. Not every operating system and execution environment makes use of that return value: Linux/Unix-based environments do, but Windows-based environments rarely do.

Typically, a program produces some output. Here is a program that writes **Hello, World!**:

```
import std;

int main()
{
      std::cout << "Hello, World!\n";
}
```

The line **import std;** instructs the compiler to make the declarations of the standard library available. Without these declarations, the expression

```
std::cout << "Hello, World!\n"
```

would make no sense. The operator **<<** ("put to") writes its second argument onto its first. In this case, the string literal **"Hello, World!\n"** is written onto the standard output stream **std::cout**. A string literal is a sequence of characters surrounded by double quotes. In a string literal, the backslash character **** followed by another character denotes a single "special character." In this case, **\n** is the newline character, so that the characters written are **Hello, World!** followed by a newline.

The **std::** specifies that the name **cout** is to be found in the standard-library namespace (§3.3). I usually leave out the **std::** when discussing standard features; §3.3 shows how to make names from a namespace visible without explicit qualification.

The **import** directive is new in C++20 and presenting all of the standard library as a module **std** is not yet standard. This will be explained in §3.2.2. If you have trouble with **import std;**, try the old-fashioned and conventional

```
#include <iostream>              // include the declarations for the I/O stream library

int main()
{
      std::cout << "Hello, World!\n";
}
```

This will be explained in §3.2.1 and has worked on all C++ implementations since 1998 (§19.1.1).

Essentially all executable code is placed in functions and called directly or indirectly from **main()**. For example:

```
import std;                      // import the declarations for the standard library

using namespace std;             // make names from std visible without std:: (§3.3)

double square(double x)          // square a double-precision floating-point number
{
      return x*x;
}

void print_square(double x)
{
      cout << "the square of " << x << " is " << square(x) << "\n";
}
```

```
int main()
{
    print_square(1.234);       // print: the square of 1.234 is 1.52276
}
```

A "return type" void indicates that a function does not return a value.

1.3 Functions（函数）

The main way of getting something done in a C++ program is to call a function to do it. Defining a function is the way you specify how an operation is to be done. A function cannot be called unless it has been declared.

A function declaration gives the name of the function, the type of the value returned (if any), and the number and types of the arguments that must be supplied in a call. For example:

```
Elem* next_elem();          // no argument; return a pointer to Elem (an Elem*)
void exit(int);             // int argument; return nothing
double sqrt(double);        // double argument; return a double
```

In a function declaration, the return type comes before the name of the function and the argument types come after the name enclosed in parentheses.

The semantics of argument passing are identical to the semantics of initialization (§3.4.1). That is, argument types are checked and implicit argument type conversion takes place when necessary (§1.4). For example:

```
double s2 = sqrt(2);             // call sqrt() with the argument double{2}
double s3 = sqrt("three");       // error: sqrt() requires an argument of type double
```

The value of such compile-time checking and type conversion should not be underestimated.

A function declaration may contain argument names. This can be a help to the reader of a program, but unless the declaration is also a function definition, the compiler simply ignores such names. For example:

```
double sqrt(double d);      // return the square root of d
double square(double);      // return the square of the argument
```

The type of a function consists of its return type followed by the sequence of its argument types in parentheses. For example:

```
double get(const vector<double>& vec, int index);      // type: double(const vector<double>&,int)
```

A function can be a member of a class (§2.3, §5.2.1). For such a *member function*, the name of its class is also part of the function type. For example:

```
char& String::operator[](int index);                   // type: char& String::(int)
```

We want our code to be comprehensible because that is the first step on the way to maintainability. The first step to comprehensibility is to break computational tasks into meaningful chunks (represented as functions and classes) and name those. Such functions then provide the basic vocabulary of computation, just as the types (built-in and user-defined) provide the basic vocabulary of data.

The C++ standard algorithms (e.g., **find**, **sort**, and **iota**) provide a good start (Chapter 13). Next, we can compose functions representing common or specialized tasks into larger computations.

The number of errors in code correlates strongly with the amount of code and the complexity of the code. Both problems can be addressed by using more and shorter functions. Using a function to do a specific task often saves us from writing a specific piece of code in the middle of other code; making it a function forces us to name the activity and document its dependencies. If we cannot find a suitable name, there is a high probability that we have a design problem.

If two functions are defined with the same name, but with different argument types, the compiler will choose the most appropriate function to invoke for each call. For example:

```
void print(int);        // takes an integer argument
void print(double);     // takes a floating-point argument
void print(string);     // takes a string argument

void user()
{
        print(42);              // calls print(int)
        print(9.65);            // calls print(double)
        print("Barcelona");     // calls print(string)
}
```

If two alternative functions could be called, but neither is better than the other, the call is deemed ambiguous and the compiler gives an error. For example:

```
void print(int,double);
void print(double,int);

void user2()
{
        print(0,0);     // error: ambiguous
}
```

Defining multiple functions with the same name is known as *function overloading* and is one of the essential parts of generic programming (§8.2). When a function is overloaded, each function of the same name should implement the same semantics. The **print()** functions are an example of this; each **print()** prints its argument.

1.4 Types, Variables, and Arithmetic（类型、变量与运算）

Every name and every expression has a type that determines the operations that may be performed on it. For example, the declaration

```
int inch;
```

specifies that **inch** is of type **int**; that is, **inch** is an integer variable.

A *declaration* is a statement that introduces an entity into the program and specifies its type:

- A *type* defines a set of possible values and a set of operations (for an object).
- An *object* is some memory that holds a value of some type.
- A *value* is a set of bits interpreted according to a type.
- A *variable* is a named object.

C++ offers a small zoo of fundamental types, but since I'm not a zoologist, I will not list them all. You can find them all in reference sources, such as the [Cppreference] on the Web. Examples are:

```
bool          // Boolean, possible values are true and false
char          // character, for example, 'a', 'z', and '9'
int           // integer, for example, -273, 42, and 1066
double        // double-precision floating-point number, for example, -273.15, 3.14, and 6.626e-34
unsigned      // non-negative integer, for example, 0, 1, and 999 (use for bitwise logical operations)
```

Each fundamental type corresponds directly to hardware facilities and has a fixed size that determines the range of values that can be stored in it:

bool:
char:
int:
double:
unsigned:

A **char** variable is of the natural size to hold a character on a given machine (typically an 8-bit byte), and the sizes of other types are multiples of the size of a **char**. The size of a type is implementation-defined (i.e., it can vary among different machines) and can be obtained by the **sizeof** operator; for example, **sizeof(char)** equals **1** and **sizeof(int)** is often **4**. When we want a type of a specific size, we use a standard-library type aliase, such as **int32_t** (§17.8).

Numbers can be floating-point or integers.

- Floating-point literals are recognized by a decimal point (e.g., **3.14**) or by an exponent (e.g., **314e–2**).
- Integer literals are by default decimal (e.g., **42** means forty-two). A **0b** prefix indicates a binary (base 2) integer literal (e.g., **0b10101010**). A **0x** prefix indicates a hexadecimal (base 16) integer literal (e.g., **0xBAD12CE3**). A **0** prefix indicates an octal (base 8) integer literal (e.g., **0334**).

To make long literals more readable for humans, we can use a single quote (') as a digit separator. For example, π is about **3.14159'26535'89793'23846'26433'83279'50288** or if you prefer hexadecimal notation **0x3.243F'6A88'85A3'08D3**.

1.4.1 Arithmetic （算术运算）

The arithmetic operators can be used for appropriate combinations of the fundamental types:

```
x+y      // plus
+x       // unary plus
x-y      // minus
-x       // unary minus
x*y      // multiply
x/y      // divide
x%y      // remainder (modulus) for integers
```

So can the comparison operators:

```
x==y     // equal
x!=y     // not equal
x<y      // less than
x>y      // greater than
x<=y     // less than or equal
x>=y     // greater than or equal
```

Furthermore, logical operators are provided:

```
x&y      // bitwise and
x|y      // bitwise or
x^y      // bitwise exclusive or
~x       // bitwise complement
x&&y     // logical and
x||y     // logical or
!x       // logical not (negation)
```

A bitwise logical operator yields a result of the operand type for which the operation has been performed on each bit. The logical operators **&&** and || simply return **true** or **false** depending on the values of their operands.

In assignments and in arithmetic operations, C++ performs all meaningful conversions between the basic types so that they can be mixed freely:

```
void some_function()        // function that doesn't return a value
{
    double d = 2.2;         // initialize floating-point number
    int i = 7;              // initialize integer
    d = d+i;                // assign sum to d
    i = d*i;                // assign product to i; beware: truncating the double d*i to an int
}
```

The conversions used in expressions are called *the usual arithmetic conversions* and aim to ensure that expressions are computed at the highest precision of their operands. For example, an addition of a **double** and an **int** is calculated using double-precision floating-point arithmetic.

Note that = is the assignment operator and == tests equality.

In addition to the conventional arithmetic and logical operators, C++ offers more specific operations for modifying a variable:

```
x+=y     // x = x+y
++x      // increment: x = x+1
x-=y     // x = x-y
--x      // decrement: x = x-1
x*=y     // scaling: x = x*y
x/=y     // scaling: x = x/y
x%=y     // x = x%y
```

These operators are concise, convenient, and very frequently used.

The order of evaluation is left-to right for `x.y`, `x->y`, `x(y)`, `x[y]`, `x<<y`, `x>>y`, `x&&y`, and `x||y`. For assignments (e.g., `x+=y`), the order is right-to-left. For historical reasons realated to optimization, the order of evaluation of other expressions (e.g., `f(x)+g(y)`) and of function arguments (e.g., `h(f(x),g(y))`) is unfortunately unspecified.

1.4.2 Initialization （初始化）

Before an object can be used, it must be given a value. C++ offers a variety of notations for expressing initialization, such as the `=` used above, and a universal form based on curly-brace-delimited initializer lists:

```
double d1 = 2.3;                 // initialize d1 to 2.3
double d2 {2.3};                 // initialize d2 to 2.3
double d3 = {2.3};               // initialize d3 to 2.3 (the = is optional with { ... })

complex<double> z = 1;           // a complex number with double-precision floating-point scalars
complex<double> z2 {d1,d2};
complex<double> z3 = {d1,d2};    // the = is optional with { ... }

vector<int> v {1, 2, 3, 4, 5, 6}; // a vector of ints
```

The `=` form is traditional and dates back to C, but if in doubt, use the general `{}`-list form. If nothing else, it saves you from conversions that lose information:

```
int i1 = 7.8;      // i1 becomes 7 (surprise?)
int i2 {7.8};      // error: floating-point to integer conversion
```

Unfortunately, conversions that lose information, *narrowing conversions*, such as `double` to `int` and `int` to `char`, are allowed and implicitly applied when you use `=` (but not when you use `{}`). The problems caused by implicit narrowing conversions are a price paid for C compatibility (§19.3).

A constant (§1.6) cannot be left uninitialized and a variable should only be left uninitialized in extremely rare circumstances. Don't introduce a name until you have a suitable value for it. User-defined types (such as `string`, `vector`, `Matrix`, `Motor_controller`, and `Orc_warrior`) can be defined to be implicitly initialized (§5.2.1).

When defining a variable, you don't need to state its type explicitly when the type can be deduced from the initializer:

```
auto b = true;      // a bool
auto ch = 'x';      // a char
auto i = 123;       // an int
```

```
auto d = 1.2;          // a double
auto z = sqrt(y);      // z has the type of whatever sqrt(y) returns
auto bb {true};        // bb is a bool
```

With **auto**, we tend to use the = because there is no potentially troublesome type conversion involved, but if you prefer to use {} initialization consistently, you can do that instead.

We use **auto** where we don't have a specific reason to mention the type explicitly. "Specific reasons" include:

- The definition is in a large scope where we want to make the type clearly visible to readers of our code.
- The type of the initializer isn't obvious.
- We want to be explicit about a variable's range or precision (e.g., **double** rather than **float**).

Using **auto**, we avoid redundancy and writing long type names. This is especially important in generic programming where the exact type of an object can be hard for the programmer to know and the type names can be quite long (§13.2).

1.5 Scope and Lifetime（作用域和生命周期）

A declaration introduces its name into a scope:

- *Local scope*: A name declared in a function (§1.3) or lambda (§7.3.2) is called a *local name*. Its scope extends from its point of declaration to the end of the block in which its declaration occurs. A *block* is delimited by a { } pair. Function argument names are considered local names.
- *Class scope*: A name is called a *member name* (or a *class member name*) if it is defined in a class (§2.2, §2.3, Chapter 5), outside any function (§1.3), lambda (§7.3.2), or **enum class** (§2.4). Its scope extends from the opening { of its enclosing declaration to the matching }.
- *Namespace scope*: A name is called a *namespace member name* if it is defined in a namespace (§3.3) outside any function, lambda (§7.3.2), class (§2.2, §2.3, Chapter 5), or **enum class** (§2.4). Its scope extends from the point of declaration to the end of its namespace.

A name not declared inside any other construct is called a *global name* and is said to be in the *global namespace*.

In addition, we can have objects without names, such as temporaries and objects created using **new** (§5.2.2). For example:

```
vector<int> vec;       // vec is global (a global vector of integers)

void fct(int arg)      // fct is global (names a global function)
                       // arg is local (names an integer argument)
{
     string motto {"Who dares wins"};   // motto is local
     auto p = new Record{"Hume"};       // p points to an unnamed Record (created by new)
     // ...
}
```

```
struct Record {
    string name;    // name is a member of Record (a string member)
    // ...
};
```

An object must be constructed (initialized) before it is used and will be destroyed at the end of its scope. For a namespace object the point of destruction is the end of the program. For a member, the point of destruction is determined by the point of destruction of the object of which it is a member. An object created by **new** "lives" until destroyed by **delete** (§5.2.2).

1.6 Constants（常量）

C++ supports two notions of *immutability* (an object with an unchangeable state):

- **const**: meaning roughly "I promise not to change this value." This is used primarily to specify interfaces so that data can be passed to functions using pointers and references without fear of it being modified. The compiler enforces the promise made by **const**. The value of a **const** may be calculated at run time.
- **constexpr**: meaning roughly "to be evaluated at compile time." This is used primarily to specify constants, to allow placement of data in read-only memory (where it is unlikely to be corrupted), and for performance. The value of a **constexpr** must be calculated by the compiler.

For example:

```
constexpr int dmv = 17;                   // dmv is a named constant
int var = 17;                             // var is not a constant
const double sqv = sqrt(var);             // sqv is a named constant, possibly computed at run time

double sum(const vector<double>&);        // sum will not modify its argument (§1.7)

vector<double> v {1.2, 3.4, 4.5};         // v is not a constant
const double s1 = sum(v);                 // OK: sum(v) is evaluated at run time
constexpr double s2 = sum(v);             // error: sum(v) is not a constant expression
```

For a function to be usable in a *constant expression*, that is, in an expression that will be evaluated by the compiler, it must be defined **constexpr** or **consteval**. For example:

```
constexpr double square(double x) { return x*x; }

constexpr double max1 = 1.4*square(17);    // OK: 1.4*square(17) is a constant expression
constexpr double max2 = 1.4*square(var);   // error: var is not a constant, so square(var) is not a constant
const double max3 = 1.4*square(var);       // OK: may be evaluated at run time
```

A **constexpr** function can be used for non-constant arguments, but when that is done the result is not a constant expression. We allow a **constexpr** function to be called with non-constant-expression arguments in contexts that do not require constant expressions. That way, we don't have to define essentially the same function twice: once for constant expressions and once for variables. When we want a function to be used only for evaluation at compile time, we declare it **consteval** rather than **constexpr**. For example:

```
consteval double square2(double x) { return x*x; }

constexpr double max1 = 1.4*square2(17);        // OK: 1.4*square(17) is a constant expression
const double max3 = 1.4*square2(var);           // error: var is not a constant
```

Functions declared **constexpr** or **consteval** are C++'s version of the notion of *pure functions*. They cannot have side effects and can only use information passed to them as arguments. In particular, they cannot modify non-local variables, but they can have loops and use their own local variables. For example:

```
constexpr double nth(double x, int n)    // assume 0<=n
{
    double res = 1;
    int i = 0;
    while (i<n) {    // while-loop: do while the condition is true (§1.7.1)
        res *= x;
        ++i;
    }
    return res;
}
```

In a few places, constant expressions are required by language rules (e.g., array bounds (§1.7), case labels (§1.8), template value arguments (§7.2), and constants declared using **constexpr**). In other cases, compile-time evaluation is important for performance. Independent of performance issues, the notion of immutability (an object with an unchangeable state) is an important design concern.

1.7 Pointers, Arrays, and References（指针、数组和引用）

The most fundamental collection of data is a contiguously allocated sequence of elements of the same type, called an *array*. This is basically what the hardware offers. An array of elements of type **char** can be declared like this:

```
char v[6];          // array of 6 characters
```

Similarly, a pointer can be declared like this:

```
char* p;            // pointer to character
```

In declarations, **[ ]** means "array of" and * means "pointer to." All arrays have **0** as their lower bound, so **v** has six elements, **v[0]** to **v[5]**. The size of an array must be a constant expression (§1.6). A pointer variable can hold the address of an object of the appropriate type:

```
char* p = &v[3];    // p points to v's fourth element
char x = *p;        // *p is the object that p points to
```

In an expression, prefix unary * means "contents of" and prefix unary **&** means "address of." We can represent that graphically like this:

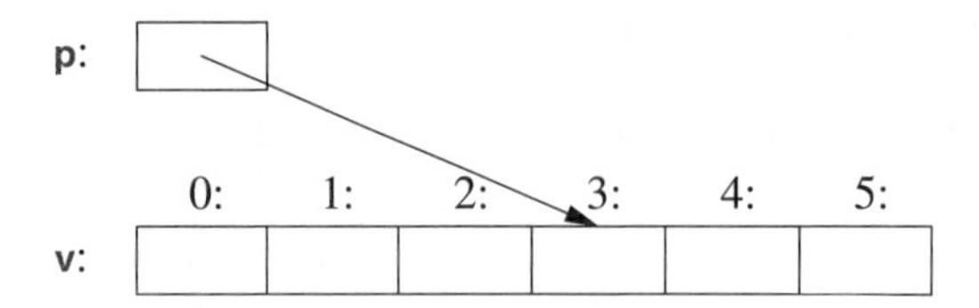

Consider printing the elements of an array:

```
void print()
{
      int v1[10] = {0, 1, 2, 3, 4, 5, 6, 7, 8, 9};

      for (auto i=0; i!=10; ++i)  // print elements
            cout << v[i] << '\n';
      // ...
}
```

This **for**-statement can be read as "set **i** to zero; while **i** is not **10**, print the **i**th element and increment **i**." C++ also offers a simpler **for**-statement, called a range-**for**-statement, for loops that traverse a sequence in the simplest way:

```
void print2()
{
      int v[] = {0, 1, 2, 3, 4, 5, 6, 7, 8, 9};

      for (auto x : v)                              // for each x in v
            cout << x << '\n';

      for (auto x : {10, 21, 32, 43, 54, 65}) // for each integer in the list
            cout << x << '\n';
      // ...
}
```

The first range-**for**-statement can be read as "for every element of **v**, from the first to the last, place a copy in **x** and print it." Note that we don't have to specify an array bound when we initialize it with a list. The range-**for**-statement can be used for any sequence of elements (§13.1).

If we didn't want to copy the values from **v** into the variable **x**, but rather just have **x** refer to an element, we could write:

```
void increment()
{
      int v[] = {0, 1, 2, 3, 4, 5, 6, 7, 8, 9};

      for (auto& x : v)       // add 1 to each x in v
            ++x;
      // ...
}
```

In a declaration, the unary suffix **&** means "reference to." A reference is similar to a pointer, except that you don't need to use a prefix $*$ to access the value referred to by the reference. Also, a reference cannot be made to refer to a different object after its initialization.

References are particularly useful for specifying function arguments. For example:

```
void sort(vector<double>& v);        // sort v (v is a vector of doubles)
```

By using a reference, we ensure that for a call **sort(my_vec)**, we do not copy **my_vec**. Therefore, it really is **my_vec** that is sorted and not a copy of it.

When we don't want to modify an argument but still don't want the cost of copying, we use a **const** reference (§1.6); that is, a reference to a **const**. For example:

```
double sum(const vector<double>&)
```

Functions taking **const** references are very common.

When used in declarations, operators (such as **&**, *****, and **[ ]**) are called *declarator operators*:

```
T a[n]     // T[n]: a is an array of n Ts
T* p       // T*: p is a pointer to T
T& r       // T&: r is a reference to T
T f(A)     // T(A): f is a function taking an argument of type A returning a result of type T
```

1.7.1 The Null Pointer （空指针）

We try to ensure that a pointer always points to an object so that dereferencing it is valid. When we don't have an object to point to or if we need to represent the notion of "no object available" (e.g., for an end of a list), we give the pointer the value **nullptr** ("the null pointer"). There is only one **nullptr** shared by all pointer types:

```
double* pd = nullptr;
Link<Record>* lst = nullptr;    // pointer to a Link to a Record
int x = nullptr;                // error: nullptr is a pointer not an integer
```

It is often wise to check that a pointer argument actually points to something:

```
int count_x(const char* p, char x)
     // count the number of occurrences of x in p[]
     // p is assumed to point to a zero-terminated array of char (or to nothing)
{
     if (p==nullptr)
          return 0;
     int count = 0;
     for (; *p!=0; ++p)
          if (*p==x)
               ++count;
     return count;
}
```

We can advance a pointer to point to the next element of an array using **++** and also leave out the initializer in a **for**-statement if we don't need it.

The definition of **count_x()** assumes that the **char*** is a *C-style string*, that is, that the pointer points to a zero-terminated array of **char**. The characters in a string literal are immutable, so to handle **count_x("Hello!")**, I declared **count_x()** a **const char*** argument.

In older code, **0** or **NULL** is typically used instead of **nullptr**. However, using **nullptr** eliminates potential confusion between integers (such as **0** or **NULL**) and pointers (such as **nullptr**).

In the **count_x()** example, we are not using the initializer part of the **for**-statement, so we can use the simpler **while**-statement:

```
int count_x(const char* p, char x)
    // count the number of occurrences of x in p[]
    // p is assumed to point to a zero-terminated array of char (or to nothing)
{
    if (p==nullptr)
        return 0;
    int count = 0;
    while (*p) {
        if (*p==x)
            ++count;
        ++p;
    }
    return count;
}
```

The **while**-statement executes until its condition becomes **false**.

A test of a numeric value (e.g., **while (*p)** in **count_x()**) is equivalent to comparing the value to **0** (e.g., **while (*p!=0)**). A test of a pointer value (e.g., **if (p)**) is equivalent to comparing the value to **nullptr** (e.g., **if (p!=nullptr)**).

There is no "null reference." A reference must refer to a valid object (and implementations assume that it does). There are obscure and clever ways to violate that rule; don't do that.

1.8 Tests（检验）

C++ provides a conventional set of statements for expressing selection and looping, such as **if**-statements, **switch**-statements, **while**-loops, and **for**-loops. For example, here is a simple function that prompts the user and returns a Boolean indicating the response:

```
bool accept()
{
    cout << "Do you want to proceed (y or n)?\n";    // write question
    char answer = 0;                                 // initialize to a value that will not appear on input
    cin >> answer;                                   // read answer

    if (answer == 'y')
        return true;
    return false;
}
```

To match the **<<** output operator ("put to"), the **>>** operator ("get from") is used for input; **cin** is the standard input stream (Chapter 11). The type of the right-hand operand of **>>** determines what input is accepted, and its right-hand operand is the target of the input operation. The **\n** character at the end of the output string represents a newline (§1.2.1).

Note that the definition of **answer** appears where it is needed (and not before that). A declaration can appear anywhere a statement can.

The example could be improved by taking an **n** (for "no") answer into account:

```
bool accept2()
{
    cout << "Do you want to proceed (y or n)?\n";    // write question
    char answer = 0;                                 // initialize to a value that will not appear on input
    cin >> answer;                                   // read answer

    switch (answer) {
    case 'y':
        return true;
    case 'n':
        return false;
    default:
        cout << "I'll take that for a no.\n";
        return false;
    }
}
```

A **switch**-statement tests a value against a set of constants. Those constants, called **case**-labels, must be distinct, and if the value tested does not match any of them, the **default** is chosen. If the value doesn't match any **case**-label and no **default** is provided, no action is taken.

We don't have to exit a **case** by returning from the function that contains its **switch**-statement. Often, we just want to continue execution with the statement following the **switch**-statement. We can do that using a **break** statement. As an example, consider an overly clever, yet primitive, parser for a trivial command video game:

```
void action()
{
    while (true) {
        cout << "enter action:\n";          // request action
        string act;
        cin >> act;                         // read characters into a string
        Point delta {0,0};                  // Point holds an {x,y} pair

        for (char ch : act) {
            switch (ch) {
            case 'u':   // up
            case 'n':   // north
                ++delta.y;
                break;
            case 'r':   // right
            case 'e':   // east
                ++delta.x;
                break;
            // ... more actions ...
```

```
                default:
                    cout << "I freeze!\n";
                }
                move(current+delta*scale);
                update_display();
            }
        }
    }
```

Like a for-statement (§1.7), an if-statement can introduce a variable and test it. For example:

```
void do_something(vector<int>& v)
{
    if (auto n = v.size(); n!=0) {
        // ... we get here if n!=0 ...
    }
    // ...
}
```

Here, the integer n is defined for use within the if-statement, initialized with v.size(), and immediately tested by the n!=0 condition after the semicolon. A name declared in a condition is in scope on both branches of the if-statement.

As with the for-statement, the purpose of declaring a name in the condition of an if-statement is to keep the scope of the variable limited to improve readability and minimize errors.

The most common case is testing a variable against 0 (or the nullptr). To do that, simply leave out the explicit mention of the condition. For example:

```
void do_something(vector<int>& v)
{
    if (auto n = v.size()) {
        // ... we get here if n!=0 ...
    }
    // ...
}
```

Prefer to use this terser and simpler form when you can.

1.9 Mapping to Hardware（映射到硬件）

C++ offers a direct mapping to hardware. When you use one of the fundamental operations, the implementation is what the hardware offers, typically a single machine operation. For example, adding two ints, x+y executes an integer add machine instruction.

A C++ implementation sees a machine's memory as a sequence of memory locations into which it can place (typed) objects and address them using pointers:

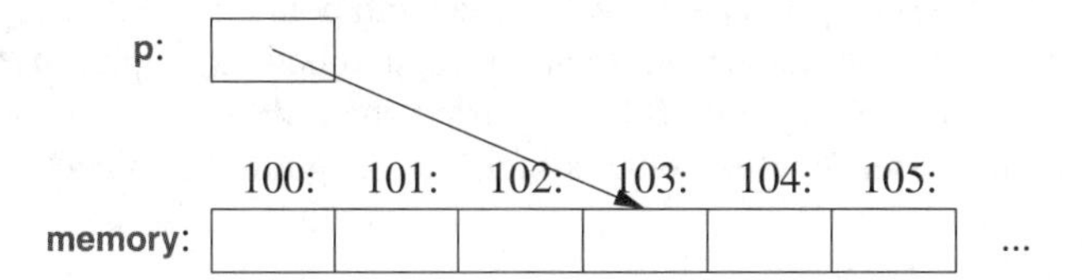

A pointer is represented in memory as a machine address, so the numeric value of **p** in this figure would be **103**. If this looks much like an array (§1.7), that's because an array is C++'s basic abstraction of "a contiguous sequence of objects in memory."

The simple mapping of fundamental language constructs to hardware is crucial for the raw low-level performance for which C and C++ have been famous for decades. The basic machine model of C and C++ is based on computer hardware, rather than some form of mathematics.

1.9.1 Assignment（赋值）

An assignment of a built-in type is a simple machine copy operation. Consider:

```
int x = 2;
int y = 3;
x = y;              // x becomes 3; so we get x==y
```

This is obvious. We can represent that graphically like this:

x:	2	y:	3	x = y;	x:	3	y:	3

The two objects are independent. We can change the value of **y** without affecting the value of **x**. For example, **x=99** will not change the value of **y**. Unlike Java, C#, and other languages, but like C, that is true for all types, not just for **int**s.

If we want different objects to refer to the same (shared) value, we must say so. For example:

```
int x = 2;
int y = 3;
int* p = &x;
int* q = &y;        // p!=q and *p!=*q
p = q;              // p becomes &y; now p==q, so (obviously)*p==*q
```

We can represent that graphically like this:

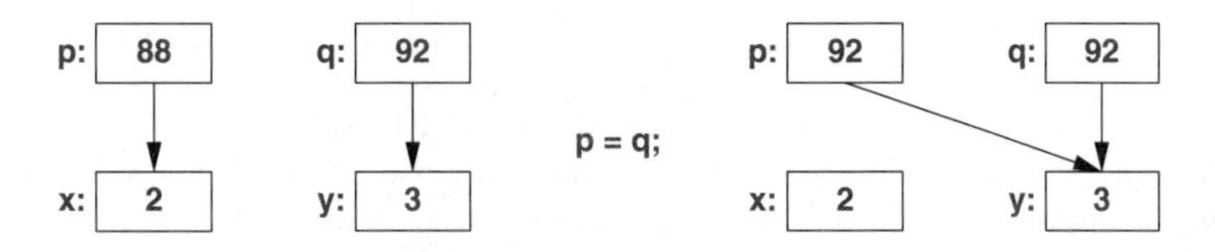

I arbitrarily chose **88** and **92** as the addresses of the **int**s. Again, we can see that the assigned-to object gets the value from the assigned object, yielding two independent objects (here, pointers), with the same value. That is, p=q gives p==q. After p=q, both pointers point to y.

A reference and a pointer both refer/point to an object and both are represented in memory as a machine address. However, the language rules for using them differ. Assignment to a reference does not change what the reference refers to but assigns to the referenced object:

```
int x = 2;
int y = 3;
int& r = x;         // r refers to x
int& r2 = y;        // r2 refers to y
r = r2;             // read through r2, write through r: x becomes 3
```

We can represent that graphically like this:

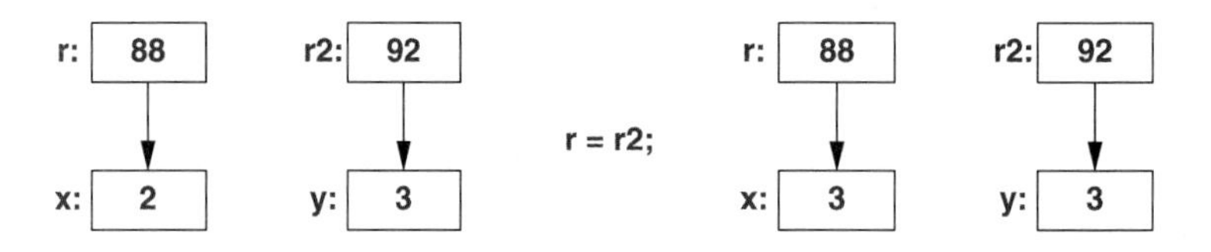

To access the value pointed to by a pointer, you use *; that is implicitly done for a reference.

After x=y, we have x==y for every built-in type and well-designed user-defined type (Chapter 2) that offers = (assignment) and == (equality comparison).

1.9.2 Initialization（初始化）

Initialization differs from assignment. In general, for an assignment to work correctly, the assigned-to object must have a value. On the other hand, the task of initialization is to make an uninitialized piece of memory into a valid object. For almost all types, the effect of reading from or writing to an uninitialized variable is undefined. Consider references:

```
int x = 7;
int& r {x};         // bind r to x (r refers to x)
r = 7;              // assign to whatever r refers to

int& r2;            // error: uninitialized reference
r2 = 99;            // assign to whatever r2 refers to
```

Fortunately, we cannot have an uninitialized reference; if we could, then that r2=99 would assign 99 to some unspecified memory location; the result would eventually lead to bad results or a crash.

You can use = to initialize a reference but please don't let that confuse you. For example:

```
int& r = x;         // bind r to x (r refers to x)
```

This is still initialization and binds r to x, rather than any form of value copy.

The distinction between initialization and assignment is also crucial to many user-defined types, such as string and vector, where an assigned-to object owns a resource that needs to eventually be released (§6.3).

The basic semantics of argument passing and function value return are that of initialization (§3.4). For example, that's how we get pass-by-reference (§3.4.1).

1.10 Advice（建议）

The advice here is a subset of the C++ Core Guidelines [Stroustrup,2015]. References to guidelines look like this [CG: ES.23], meaning the 23rd rule in the Expressions and Statement section. Generally, a core guideline offers further rationale and examples.

[1] Don't panic! All will become clear in time; §1.1; [CG: In.0].
[2] Don't use the built-in features exclusively. Many fundamental (built-in) features are usually best used indirectly through libraries, such as the ISO C++ standard library (Chapters 9–18); [CG: P.13].
[3] **#include** or (preferably) **import** the libraries needed to simplify programming; §1.2.1.
[4] You don't have to know every detail of C++ to write good programs.
[5] Focus on programming techniques, not on language features.
[6] The ISO C++ standard is the final word on language definition issues; §19.1.3; [CG: P.2].
[7] "Package" meaningful operations as carefully named functions; §1.3; [CG: F.1].
[8] A function should perform a single logical operation; §1.3 [CG: F.2].
[9] Keep functions short; §1.3; [CG: F.3].
[10] Use overloading when functions perform conceptually the same task on different types; §1.3.
[11] If a function may have to be evaluated at compile time, declare it **constexpr**; §1.6; [CG: F.4].
[12] If a function must be evaluated at compile time, declare it **consteval**; §1.6.
[13] If a function may not have side effects, declare it **constexpr** or **consteval**; §1.6; [CG: F.4].
[14] Understand how language primitives map to hardware; §1.4, §1.7, §1.9, §2.3, §5.2.2, §5.4.
[15] Use digit separators to make large literals readable; §1.4; [CG: NL.11].
[16] Avoid complicated expressions; [CG: ES.40].
[17] Avoid narrowing conversions; §1.4.2; [CG: ES.46].
[18] Minimize the scope of a variable; §1.5, §1.8.
[19] Keep scopes small; §1.5; [CG: ES.5].
[20] Avoid "magic constants"; use symbolic constants; §1.6; [CG: ES.45].
[21] Prefer immutable data; §1.6; [CG: P.10].
[22] Declare one name (only) per declaration; [CG: ES.10].
[23] Keep common and local names short; keep uncommon and nonlocal names longer; [CG: ES.7].
[24] Avoid similar-looking names; [CG: ES.8].
[25] Avoid **ALL_CAPS** names; [CG: ES.9].
[26] Prefer the **{}**-initializer syntax for declarations with a named type; §1.4; [CG: ES.23].
[27] Use **auto** to avoid repeating type names; §1.4.2; [CG: ES.11].
[28] Avoid uninitialized variables; §1.4; [CG: ES.20].
[29] Don't declare a variable until you have a value to initialize it with; §1.7, §1.8; [CG: ES.21].
[30] When declaring a variable in the condition of an **if**-statement, prefer the version with the implicit test against **0** or **nullptr**; §1.8.
[31] Prefer range-**for** loops over **for**-loops with an explicit loop variable; §1.7.
[32] Use **unsigned** for bit manipulation only; §1.4; [CG: ES.101] [CG: ES.106].
[33] Keep use of pointers simple and straightforward; §1.7; [CG: ES.42].
[34] Use **nullptr** rather than **0** or **NULL**; §1.7; [CG: ES.47].

[35] Don't say in comments what can be clearly stated in code; [CG: NL.1].
[36] State intent in comments; [CG: NL.2].
[37] Maintain a consistent indentation style; [CG: NL.4].

2

User-Defined Types

（用户自定义类型）

Don't Panic!
– Douglas Adams

- Introduction
- Structures
- Classes
- Enumerations
- Unions
- Advice

2.1 Introduction （引言）

We call the types that can be built from the fundamental types (§1.4), the **const** modifier (§1.6), and the declarator operators (§1.7) *built-in types*. C++'s set of built-in types and operations is rich, but deliberately low-level. They directly and efficiently reflect the capabilities of conventional computer hardware. However, they don't provide the programmer with high-level facilities to conveniently write advanced applications. Instead, C++ augments the built-in types and operations with a sophisticated set of *abstraction mechanisms* out of which programmers can build such high-level facilities.

The C++ abstraction mechanisms are primarily designed to let programmers design and implement their own types, with suitable representations and operations, and for programmers to simply and elegantly use such types. Types built out of other types using C++'s abstraction mechanisms are called *user-defined types*. They are referred to as *classes* and *enumerations*. User-defined types can be built out of both built-in types and other user-defined types. Most of this book is devoted to the design, implementation, and use of user-defined types. User-defined types are often preferred over built-in types because they are easier to use, less error-prone, and typically as efficient for what they do as direct use of built-in types, or even more efficient.

The rest of this chapter presents the simplest and most fundamental facilities for defining and using types. Chapters 4–8 are a more complete description of the abstraction mechanisms and the programming styles they support. User-defined types provide the backbone of the standard library, so the standard-library chapters, 9–17, provide examples of what can be built using the language facilities and programming techniques presented in Chapters 1–8.

2.2 Structures（结构）

The first step in building a new type is often to organize the elements it needs into a data structure, a **struct**:

```
struct Vector {
      double* elem; // pointer to elements
      int sz;           // number of elements
};
```

This first version of **Vector** consists of an **int** and a **double***.

A variable of type **Vector** can be defined like this:

```
Vector v;
```

However, by itself that is not of much use because **v**'s **elem** pointer doesn't point to anything. For it to be useful, we must give **v** some elements to point to. For example::

```
void vector_init(Vector& v, int s)     // initialize a Vector
{
      v.elem = new double[s];  // allocate an array of s doubles
      v.sz = s;
}
```

That is, **v**'s **elem** member gets a pointer produced by the **new** operator and **v**'s **sz** member gets the number of elements. The **&** in **Vector&** indicates that we pass **v** by non-**const** reference (§1.7); that way, **vector_init()** can modify the vector passed to it.

The **new** operator allocates memory from an area called the *free store* (also known as *dynamic memory* and *heap*). Objects allocated on the free store are independent of the scope from which they are created and "live" until they are destroyed using the **delete** operator (§5.2.2).

A simple use of **Vector** looks like this:

```
double read_and_sum(int s)
      // read s integers from cin and return their sum; s is assumed to be positive
{
      Vector v;
      vector_init(v,s);                    // allocate s elements for v

      for (int i=0; i!=s; ++i)
            cin>>v.elem[i];                // read into elements
```

```
        double sum = 0;
        for (int i=0; i!=s; ++i)
                sum+=v.elem[i];                 // compute the sum of the elements
        return sum;
}
```

There is a long way to go before our Vector is as elegant and flexible as the standard-library vector. In particular, a user of Vector has to know every detail of Vector's representation. The rest of this chapter and the next two gradually improve Vector as an example of language features and techniques. Chapter 12 presents the standard-library vector, which contains many nice improvements.

I use vector and other standard-library components as examples

- to illustrate language features and design techniques, and
- to help you learn and use the standard-library components.

Don't reinvent standard-library components such as vector and string; use them. The standard-library types have lower-case names, so to distinguish names of types used to illustrate design and implementation techniques (e.g., Vector and String), I capitalize them.

We use . (dot) to access struct members through a name (and through a reference) and -> to access struct members through a pointer. For example:

```
void f(Vector v, Vector& rv, Vector* pv)
{
        int i1 = v.sz;          // access through name
        int i2 = rv.sz;         // access through reference
        int i3 = pv->sz;        // access through pointer
}
```

2.3 Classes（类）

Having data specified separately from the operations on it has advantages, such as the ability to use the data in arbitrary ways. However, a tighter connection between the representation and the operations is needed for a user-defined type to have all the properties expected of a "real type." In particular, we often want to keep the representation inaccessible to users so as to simplify use, guarantee consistent use of the data, and allow us to later improve the representation. To do that, we have to distinguish between the interface to a type (to be used by all) and its implementation (which has access to the otherwise inaccessible data). The language mechanism for that is called a *class*. A class has a set of *members*, which can be data, function, or type members.

The interface of a class is defined by its public members, and its private members are accessible only through that interface. The public and private parts of a class declaration can appear in any order, but conventionally we place the public declarations first and the private declarations later, except when we want to emphasize the representation. For example:

```
class Vector {
public:
     Vector(int s) :elem{new double[s]}, sz{s} { }   // construct a Vector
     double& operator[](int i) { return elem[i]; }   // element access: subscripting
     int size() { return sz; }
private:
     double* elem;  // pointer to the elements
     int sz;        // the number of elements
};
```

Given that, we can define a variable of our new type **Vector**:

```
Vector v(6);     // a Vector with 6 elements
```

We can illustrate a **Vector** object graphically:

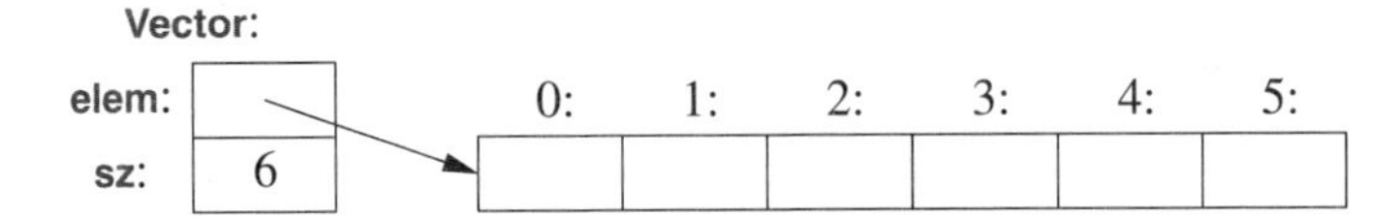

Basically, the **Vector** object is a "handle" containing a pointer to the elements (**elem**) and the number of elements (**sz**). The number of elements (6 in the example) can vary from **Vector** object to **Vector** object, and a **Vector** object can have a different number of elements at different times (§5.2.3). However, the **Vector** object itself is always the same size. This is the basic technique for handling varying amounts of information in C++: a fixed-size handle referring to a variable amount of data "elsewhere" (e.g., on the free store allocated by **new**; §5.2.2). How to design and use such objects is the main topic of Chapter 5.

Here, the representation of a **Vector** (the members **elem** and **sz**) is accessible only through the interface provided by the **public** members: **Vector()**, **operator[]()**, and **size()**. The **read_and_sum()** example from §2.2 simplifies to:

```
double read_and_sum(int s)
{
     Vector v(s);                          // make a vector of s elements
     for (int i=0; i!=v.size(); ++i)
          cin>>v[i];                       // read into elements

     double sum = 0;
     for (int i=0; i!=v.size(); ++i)
          sum+=v[i];                       // take the sum of the elements
     return sum;
}
```

A member function with the same name as its class is called a *constructor*, that is, a function used to *construct* objects of a class. So, the constructor, **Vector()**, replaces **vector_init()** from §2.2. Unlike an ordinary function, a constructor is guaranteed to be used to initialize objects of its class. Thus, defining a constructor eliminates the problem of uninitialized variables for a class.

Vector(int) defines how objects of type Vector are constructed. In particular, it states that it needs an integer to do that. That integer is used as the number of elements. The constructor initializes the Vector members using a member initializer list:

```
:elem{new double[s]}, sz{s}
```

That is, we first initialize elem with a pointer to s elements of type double obtained from the free store. Then, we initialize sz to s.

Access to elements is provided by a subscript function, called operator[]. It returns a reference to the appropriate element (a double& allowing both reading and writing).

The size() function is supplied to give users the number of elements.

Obviously, error handling is completely missing, but we'll return to that in Chapter 4. Similarly, we did not provide a mechanism to "give back" the array of doubles acquired by new; §5.2.2 shows how to define a destructor to elegantly do that.

There is no fundamental difference between a struct and a class; a struct is simply a class with members public by default. For example, you can define constructors and other member functions for a struct.

2.4 Enumerations（枚举）

In addition to classes, C++ supports a simple form of user-defined type for which we can enumerate the values:

```
enum class Color { red, blue, green };
enum class Traffic_light { green, yellow, red };

Color col = Color::red;
Traffic_light light = Traffic_light::red;
```

Note that enumerators (e.g., red) are in the scope of their enum class, so that they can be used repeatedly in different enum classes without confusion. For example, Color::red is Color's red which is different from Traffic_light::red.

Enumerations are used to represent small sets of integer values. They are used to make code more readable and less error-prone than it would have been had the symbolic (and mnemonic) enumerator names not been used.

The class after the enum specifies that an enumeration is strongly typed and that its enumerators are scoped. Being separate types, enum classes help prevent accidental misuses of constants. In particular, we cannot mix Traffic_light and Color values:

```
Color x1 = red;                    // error: which red?
Color y2 = Traffic_light::red;     // error: that red is not a Color
Color z3 = Color::red;             // OK
auto x4 = Color::red;              // OK: Color::red is a Color
```

Similarly, we cannot implicitly mix Color and integer values:

```
int i = Color::red;          // error: Color::red is not an int
Color c = 2;                 // initialization error: 2 is not a Color
```

Catching attempted conversions to an enum is a good defense against errors, but often we want to initialize an enum with a value from its underlying type (by default, that's int), so that's allowed, as is explicit conversion from the underlying type:

```
Color x = Color{5};  // OK, but verbose
Color y {6};         // also OK
```

Similarly, we can explicitly convert an enum value to its underlying type:

```
int x = int(Color::red);
```

By default, an enum class has assignment, initialization, and comparisons (e.g., == and <; §1.4) defined, annd only those. However, an enumeration is a user-defined type, so we can define operators for it (§6.4):

```
Traffic_light& operator++(Traffic_light& t)          // prefix increment: ++
{
     switch (t) {
     case Traffic_light::green:     return t=Traffic_light::yellow;
     case Traffic_light::yellow:    return t=Traffic_light::red;
     case Traffic_light::red:       return t=Traffic_light::green;
     }
}

auto signal = Traffic_light::red;
Traffic_light next = ++signal;          // next becomes Traffic_light::green
```

If the repetition of the enumeration name, Traffic_light, becomes too tedious, we can abbreviate it in a scope:

```
Traffic_light& operator++(Traffic_light& t)          // prefix increment: ++
{
     using enum Traffic_light;          // here, we are using Traffic_light

     switch (t) {
     case green:    return t=yellow;
     case yellow:   return t=red;
     case red:      return t=green;
     }
}
```

If you don't ever want to explicitly qualify enumerator names and want enumerator values to be ints (without the need for an explicit conversion), you can remove the class from enum class to get a "plain" enum. The enumerators from a "plain" enum are entered into the same scope as the name of their enum and implicitly convert to their integer values. For example:

```
enum Color { red, green, blue };
int col = green;
```

Here col gets the value 1. By default, the integer values of enumerators start with 0 and increase by one for each additional enumerator. The "plain" **enum**s have been in C++ (and C) since the earliest days, so even though they are less well behaved, they are common in current code.

2.5 Unions（联合）

A **union** is a **struct** in which all members are allocated at the same address so that the **union** occupies only as much space as its largest member. Naturally, a **union** can hold a value for only one member at a time. For example, consider a symbol table entry that holds a name and a value. The value can either be a **Node*** or an **int**:

```
enum class Type { ptr, num }; // a Type can hold values ptr and num (§2.4)

struct Entry {
	string name;	// string is a standard-library type
	Type t;
	Node* p; // use p if t==Type::ptr
	int i;		// use i if t==Type::num
};

void f(Entry* pe)
{
	if (pe->t == Type::num)
		cout << pe->i;
	// ...
}
```

The members **p** and **i** are never used at the same time, so space is wasted. It can be easily recovered by specifying that both should be members of a **union**, like this:

```
union Value {
	Node* p;
	int i;
};
```

Now **Value::p** and **Value::i** are placed at the same address of memory of each **Value** object.

This kind of space optimization can be important for applications that hold large amounts of memory so that compact representation is critical.

The language doesn't keep track of which kind of value is held by a **union**, so the programmer must do that:

```
struct Entry {
	string name;
	Type t;
	Value v;	// use v.p if t==Type::ptr; use v.i if t==Type::num
};
```

```
void f(Entry* pe)
{
      if (pe->t == Type::num)
            cout << pe->v.i;
      // ...
}
```

Maintaining the correspondence between a *type field*, sometimes called a *discriminant* or a *tag*, (here, **t**) and the type held in a **union** is error-prone. To avoid errors, we can enforce that correspondence by encapsulating the union and the type field in a class and offer access only through member functions that use the union correctly. At the application level, abstractions relying on such *tagged unions* are common and useful. The use of "naked" **unions** is best minimized.

The standard library type, **variant**, can be used to eliminate most direct uses of unions. A **variant** stores a value of one of a set of alternative types (§15.4.1). For example, a **variant<Node*,int>** can hold either a **Node*** or an **int**. Using **variant**, the **Entry** example could be written as:

```
struct Entry {
      string name;
      variant<Node*,int> v;
};

void f(Entry* pe)
{
      if (holds_alternative<int>(pe->v))    // does *pe hold an int? (see §15.4.1)
            cout << get<int>(pe->v);        // get the int
      // ...
}
```

For many uses, a **variant** is simpler and safer to use than a **union**.

2.6 Advice（建议）

[1] Prefer well-defined user-defined types over built-in types when the built-in types are too low-level; §2.1.
[2] Organize related data into structures (**struct**s or **class**es); §2.2; [CG: C.1].
[3] Represent the distinction between an interface and an implementation using a **class**; §2.3; [CG: C.3].
[4] A **struct** is simply a **class** with its members **public** by default; §2.3.
[5] Define constructors to guarantee and simplify initialization of **class**es; §2.3; [CG: C.2].
[6] Use enumerations to represent sets of named constants; §2.4; [CG: Enum.2].
[7] Prefer **class enum**s over "plain" **enum**s to minimize surprises; §2.4; [CG: Enum.3].
[8] Define operations on enumerations for safe and simple use; §2.4; [CG: Enum.4].
[9] Avoid "naked" **union**s; wrap them in a class together with a type field; §2.5; [CG: C.181].
[10] Prefer **std::variant** to "naked **union**s."; §2.5.

3

Modularity

（模块化）

Mind your own business.
– Traditional

3.1 Introduction （引言）

A C++ program consists of many separately developed parts, such as functions (§1.2.1), user-defined types (Chapter 2), class hierarchies (§5.5), and templates (Chapter 7). The key to managing such a multitude of parts is to clearly define the interactions among those parts. The first and most important step is to distinguish between the interface to a part and its implementation. At the language level, C++ represents interfaces by declarations. A *declaration* specifies all that's needed to use a function or a type. For example:

```
double sqrt(double);        // the square root function takes a double and returns a double

class Vector {              // what is needed to use a Vector
public:
        Vector(int s);
        double& operator[](int i);
        int size();
```

```
private:
        double* elem; // elem points to an array of sz doubles
        int sz;
};
```

The key point here is that the function bodies, the function *definitions*, can be "elsewhere." For this example, we might like for the representation of **Vector** to be "elsewhere" also, but we will deal with that later (abstract types; §5.3). The definition of **sqrt()** will look like this:

```
double sqrt(double d)          // definition of sqrt()
{
        // ... algorithm as found in math textbook ...
}
```

For **Vector**, we need to define all three member functions:

```
Vector::Vector(int s)                        // definition of the constructor
        :elem{new double[s]}, sz{s}          // initialize members
{
}

double& Vector::operator[](int i)            // definition of subscripting
{
        return elem[i];
}

int Vector::size()                           // definition of size()
{
        return sz;
}
```

We must define **Vector**'s functions, but not **sqrt()** because it is part of the standard library. However, that makes no real difference: a library is simply "some other code we happen to use" written with the same language facilities we use.

There can be many declarations for an entity, such as a function, but only one definition.

3.2 Separate Compilation（分离编译）

C++ supports a notion of separate compilation where user code sees only declarations of the types and functions used. This can be done in two ways:

- *Header files* (§3.2.1): Place declarations in separate files, called *header files*, and textually **#include** a header file where its declarations are needed.
- *Modules* (§3.2.2): Define **module** files, compile them separately, and **import** them where needed. Only explicitly **exported** declarations are seen by code **importing** the **module**.

Either can be used to organize a program into a set of semi-independent code fragments. Such separation can be used to minimize compilation times and to enforce separation of logically distinct parts of a program (thus minimizing the chance of errors). A library is often a collection of separately compiled code fragments (e.g., functions).

The header-file technique for organizing code goes back to the earliest days of C and is still by far the most common. The use of modules is new in C++20 and offers massive advantages in code hygiene and compile time.

3.2.1 Header Files（头文件）

Traditionally, we place declarations that specify the interface to a piece of code we consider a module in a file with a name indicating its intended use. For example:

```
// Vector.h:

class Vector {
public:
	Vector(int s);
	double& operator[](int i);
	int size();
private:
	double* elem;		// elem points to an array of sz doubles
	int sz;
};
```

This declaration would be placed in a file **Vector.h**. Users then **#include** that file, called a *header file*, to access that interface. For example:

```
// user.cpp:

#include "Vector.h"		// get Vector's interface
#include <cmath>		// get the standard-library math function interface including sqrt()

double sqrt_sum(const Vector& v)
{
	double sum = 0;
	for (int i=0; i!=v.size(); ++i)
		sum+=std::sqrt(v[i]);			// sum of square roots
	return sum;
}
```

To help the compiler ensure consistency, the **.cpp** file providing the implementation of **Vector** will also include the **.h** file providing its interface:

```
// Vector.cpp:

#include "Vector.h"		// get Vector's interface

Vector::Vector(int s)
	:elem{new double[s]}, sz{s}		// initialize members
{
}
```

```
double& Vector::operator[](int i)
{
    return elem[i];
}

int Vector::size()
{
    return sz;
}
```

The code in **user.cpp** and **Vector.cpp** shares the **Vector** interface information presented in **Vector.h**, but the two files are otherwise independent and can be separately compiled. Graphically, the program fragments can be represented like this:

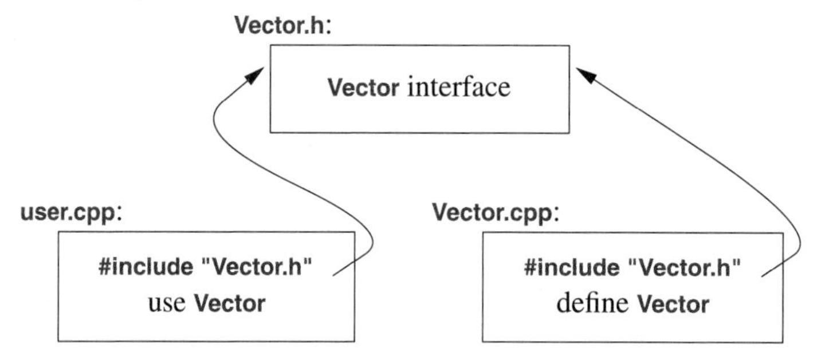

The best approach to program organization is to think of a program as a set of modules with well-defined dependencies. Header files represent that modularity through files and then exploit that modularity through separate compilation.

A **.cpp** file that is compiled by itself (including the **h** files it **#includes**) is called a *translation unit*. A program can consist of thousands of translation units.

The use of header files and **#include** is a very old way of simulating modularity with significant disadvantages:

- *Compilation time*: If you **#include header.h** in 101 translation units, the text of **header.h** will be processed by the compiler 101 times.
- *Order dependencies*: If we **#include header1.h** before **header2.h** the declarations and macros (§19.3.2.1) in **header1.h** might affect the meaning of the code in **header2.h**. If instead you **#include header2.h** before **header1.h**, it is **header2.h** that might affect the code in **header1.h**.
- *Inconsistencies*: Defining an entity, such as a type or a function, in one file and then defining it slightly differently in another file, can lead to crashes or subtle errors. This can happen if we – accidentally or deliberately – declare an entity separately in two source files, rather than putting it in a header, or through order dependencies between different header files.
- *Transitivity*: All code that is needed to express a declaration in a header file must be present in that header file. This leads to massive code bloat as header files **#include** other headers and this results in the user of a header file – accidentially or deliberately – becoming dependent on such implementation details.

Obviously, this is not ideal, and this technique has been a major source of cost and bugs since it was first introduced into C in the early 1970s. However, use of header files has been viable for decades and old code using **#include** will "live" for a very long time because it can be costly and time consuming to update large programs.

3.2.2 Modules （模块）

In C++20, we finally have a language-supported way of directly expressing modularity (§19.2.4). Consider how to express the **Vector** and **sqrt_sum()** example from §3.2 using **modules**:

```
export module Vector;    // defining the module called "Vector"

export class Vector {
public:
        Vector(int s);
        double& operator[](int i);
        int size();
private:
        double* elem;        // elem points to an array of sz doubles
        int sz;
};

Vector::Vector(int s)
        :elem{new double[s]}, sz{s}            // initialize members
{
}

double& Vector::operator[](int i)
{
        return elem[i];
}

int Vector::size()
{
        return sz;
}

export bool operator==(const Vector& v1, const Vector& v2)
{
        if (v1.size()!=v2.size())
                return false;
        for (int i = 0; i<v1.size(); ++i)
                if (v1[i]!=v2[i])
                        return false;
        return true;
}
```

This defines a **module** called **Vector**, which exports the class **Vector**, all its member functions, and the non-member function defining operator **==**.

The way we use this **module** is to **import** it where we need it. For example:

```
// file user.cpp:

import Vector;              // get Vector's interface
#include <cmath>            // get the standard-library math function interface including sqrt()

double sqrt_sum(Vector& v)
{
      double sum = 0;
      for (int i=0; i!=v.size(); ++i)
            sum+=std::sqrt(v[i]);          // sum of square roots
      return sum;
}
```

I could have **import**ed the standard-library mathematical functions also, but I used the old-fashioned **#include** just to show that we can mix old and new. Such mixing is essential for gradually upgrading older code from using **#include** to using **import**.

The differences between headers and modules are not just syntactic.

- A module is compiled once only (rather than in each translation unit in which it is used).
- Two modules can be **import**ed in either order without changing their meaning.
- If you **import** or **#include** something into a module, users of your module do not implicitly gain access to (and are not bothered by) that: **import** is not transitive.

The effects on maintainability and compile-time performance can be spectacular. For example, I have measured the "Hello, World!" program using

```
import std;
```

to compile 10 times faster than the version using

```
#include<iostream>
```

This is achieved despite the fact that module **std** contains the whole standard library, more than 10 times as much information as the **<iostream>** header. The reason is that modules only export interfaces whereas a header delivers all that it directly or indirectly contains to the compiler. This allows us to use large modules so that we don't have to remember which of a bewildering collection of headers (§9.3.4) to **#include**. From here on, I will assume **import std** for all examples.

Unfortunately, **module std** is not part of C++20. Appendix A explains how to get a **module std** if a standard-library implementation doesn't yet supply it.

When defining a module, we do not have to separate declarations and definitions into separate files; we can if that improves our source code organization, but we don't have to. We could define the simple **Vector** module like this:

```
export module Vector;     // defining the module called "Vector"

export class Vector {
      // ...
};
```

```
export bool operator==(const Vector& v1, const Vector& v2)
{
      // ...
}
```

Graphically, the program fragments can be represented like this:

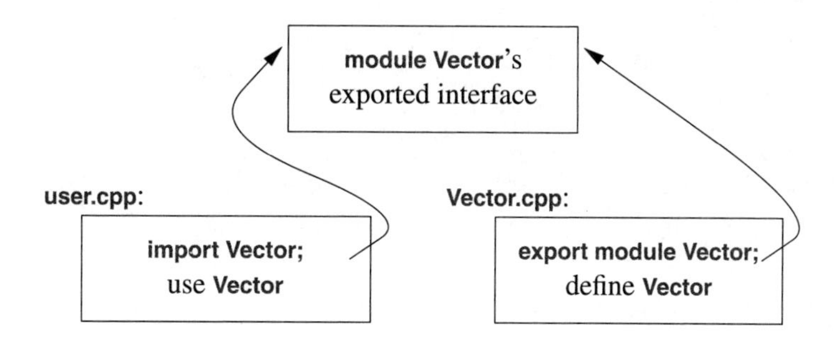

The compiler separates a module's interface, specified by the **export** specifiers, from its implementation details. Thus, the **Vector** interface is generated by the compiler and never explicitly named by the user.

Using **modules**, we don't have to complicate our code to hide implementation details from users; a **module** will only grant access to **export**ed declarations. Consider:

```
export module vector_printer;

import std;

export
template<typename T>
void print(std::vector<T>& v)  // this is the (only) function seen by users
{
      cout << "{\n";
      for (const T& val : v)
            std::cout << "   " << val << '\n';
      cout << '}';
}
```

By **import**ing this trivial module, we don't suddenly gain access to all of the standard library.

The **template<typename T>** is how we parameterize a function with a type (§7.2).

3.3 Namespaces （命名空间）

In addition to functions (§1.3), classes (§2.3), and enumerations (§2.4), C++ offers *namespaces* as a mechanism for expressing that some declarations belong together and that their names shouldn't clash with other names. For example, I might want to experiment with my own complex number type (§5.2.1, §17.4):

```
namespace My_code {
    class complex {
        // ...
    };

    complex sqrt(complex);
    // ...

    int main();
}

int My_code::main()
{
    complex z {1,2};
    auto z2 = sqrt(z);
    std::cout << '{' << z2.real() << ',' << z2.imag() << "}\n";
    // ...
}

int main()
{
    return My_code::main();
}
```

By putting my code into the namespace **My_code**, I make sure that my names do not conflict with the standard-library names in namespace **std** (§3.3). That precaution is wise, because the standard library does provide support for **complex** arithmetic (§5.2.1, §17.4).

The simplest way to access a name in another namespace is to qualify it with the namespace name (e.g., **std::cout** and **My_code::main**). The "real **main()**" is defined in the global namespace, that is, not local to a defined namespace, class, or function.

If repeatedly qualifying a name becomes tedious or distracting, we can bring the name into a scope with a *using-declaration*:

```
void my_code(vector<int>& x, vector<int>& y)
{
    using std::swap;        // make the standard-library swap available locally
    // ...
    swap(x,y);              // std::swap()
    other::swap(x,y);       // some other swap()
    // ...
}
```

A **using**-declaration makes a name from a namespace usable as if it were declared in the scope in which it appears. After **using std::swap**, it is exactly as if **swap** had been declared in **my_code()**.

To gain access to all names in the standard-library namespace, we can use a *using-directive*:

```
using namespace std;
```

A **using**-directive makes unqualified names from the named namespace accessible from the scope in which we placed the directive. So after the **using**-directive for **std**, we can simply write **cout** rather than **std::cout**. For example, we could avoid the repeated **std::** qualifications in our trivial **vector_printer** module:

```
export module vector_printer;

import std;
using namespace std;

export
template<typename T>
void print(vector<T>& v) // this is the (only) function seen by users
{
	cout << "{\n";
	for (const T& val : v)
		cout << "  " << val << '\n';
	cout << '}';
}
```

Importantly, that use of the namespace directive does not affect users of our modules; it is an implementation detail, local to the module.

By using a **using**-directive, we lose the ability to selectively use names from that namespace, so this facility should be used carefully, usually for a library that's pervasive in an application (e.g., **std**) or during a transition for an application that didn't use **namespaces**.

Namespaces are primarily used to organize larger program components, such as libraries. They simplify the composition of a program out of separately developed parts.

3.4 Function Arguments and Return Values（函数参数与返回值）

The primary and recommended way of passing information from one part of a program to another is through a function call. Information needed to perform a task is passed as arguments to a function and the results produced are passed back as return values. For example:

```
int sum(const vector<int>& v)
{
	int s = 0;
	for (const int i : v)
		s += i;
	return s;
}

vector fib = {1, 2, 3, 5, 8, 13, 21};

int x = sum(fib);		// x becomes 53
```

There are other paths through which information can be passed between functions, such as global variables (§1.5) and shared state in a class object (Chapter 5). Global variables are strongly discouraged as a known source of errors, and state should typically be shared only between functions jointly implementing a well-defined abstraction (e.g., member functions of a class; §2.3).

Given the importance of passing information to and from functions, it is not surprising that there are a variety of ways of doing it. Key concerns are:

- Is an object copied or shared?
- If an object is shared, is it mutable?
- Is an object moved, leaving an "empty object" behind (§6.2.2)?

The default behavior for both argument passing and value return is "make a copy" (§1.9), but many copies can implicitly be optimized to moves.

In the **sum()** example, the resulting **int** is copied out of **sum()** but it would be inefficient and pointless to copy the potentially very large **vector** into **sum()**, so the argument is passed by reference (indicated by the **&**; §1.7).

The **sum()** has no reason to modify its argument. This immutability is indicated by declaring the **vector** argument **const** (§1.6), so the **vector** is passed by **const**-reference.

3.4.1 Argument Passing（参数传递）

First consider how to get values into a function. By default we copy ("pass-by-value") and if we want to refer to an object in the caller's environment, we use a reference ("pass-by-reference"). For example:

```
void test(vector<int> v, vector<int>& rv)        // v is passed by value; rv is passed by reference
{
      v[1] = 99;        // modify v (a local variable)
      rv[2] = 66;       // modify whatever rv refers to
}

int main()
{
      vector fib = {1, 2, 3, 5, 8, 13, 21};
      test(fib,fib);
      cout << fib[1] << ' ' << fib[2] << '\n';        // prints 2 66
}
```

When we care about performance, we usually pass small values by-value and larger ones by-reference. Here "small" means "something that's really cheap to copy." Exactly what "small" means depends on machine architecture, but "the size of two or three pointers or less" is a good rule of thumb. If it might be significant to your performance, measure.

If we want to pass by reference for performance reasons but don't need to modify the argument, we pass-by-**const**-reference as in the **sum()** example. This is by far the most common case in ordinary good code: it is fast and not error-prone.

It is not uncommon for a function argument to have a default value; that is, a value that is considered preferred or just the most common. We can specify such a default by a *default function argument*. For example:

```
void print(int value, int base =10);   // print value in base "base"

print(x,16);      // hexadecimal
print(x,60);      // sexagesimal (Sumerian)
print(x);         // use the default: decimal
```

This is a notationally simpler alternative to overloading:

```
void print(int value, int base);        // print value in base "base"

void print(int value)                   // print value in base 10
{
     print(value,10);
}
```

Using default arguments means that there is only one definition of the function. That's usually good for comprehension and code size. When we need different code to implement the same semantics for different types, we can use overloading.

3.4.2 Value Return （返回值）

Once we have computed a result, we need to get it out of the function and back to the caller. Again, the default for value return is to copy and for small objects that's ideal. We return "by reference" only when we want to grant a caller access to something that is not local to the function. For example, a Vector can grant a user access to an element through subscripting:

```
class Vector {
public:
     // ...
     double& operator[](int i) { return elem[i]; }      // return reference to ith element
private:
     double* elem;        // elem points to an array of sz
     // ...
};
```

The ith element of a Vector exists independently of the call of the subscript operator, so we can return a reference to it.

On the other hand, a local variable disappears when the function returns, so we should not return a pointer or reference to it:

```
int& bad()
{
     int x;
     // ...
     return x;  // bad: return a reference to the local variable x
}
```

Fortunately, all major C++ compilers will catch the obvious error in bad().

Returning a reference or a value of a "small" type is efficient, but how do we pass large amounts of information out of a function? Consider:

```
Matrix operator+(const Matrix& x, const Matrix& y)
{
     Matrix res;
     // ... for all res[i,j], res[i,j] = x[i,j]+y[i,j] ...
     return res;
}
```

```
Matrix m1, m2;
// ...
Matrix m3 = m1+m2;        // no copy
```

A **Matrix** may be *very* large and expensive to copy even on modern hardware. So we don't copy, we give **Matrix** a move constructor (§6.2.2) that very cheaply moves the **Matrix** out of **operator+()**. Even if we don't define a move constructor, the compiler is often able to optimize away the copy (elide the copy) and construct the **Matrix** exactly where it is needed. This is called *copy elision*.

We should *not* regress to use manual memory management:

```
Matrix* add(const Matrix& x, const Matrix& y)      // complicated and error-prone 20th century style
{
      Matrix* p = new Matrix;
      // ... for all *p[i,j], *p[i,j] = x[i,j]+y[i,j] ...
      return p;
}

Matrix m1, m2;
// ...
Matrix* m3 = add(m1,m2);        // just copy a pointer
// ...
delete m3;                      // easily forgotten
```

Unfortunately, returning a large object by returning a pointer to it is common in older code and a major source of hard-to-find errors. Don't write such code. The **Matrix operator+()** is at least as efficient as the **Matrix add()**, but far easier to define, easier to use, and less error-prone.

If a function cannot perform its required task, it can throw an exception (§4.2). This can help avoid code from being littered with error-code tests for "exceptional problems."

3.4.3 Return Type Deduction（返回类型推导）

The return type of a function can be deduced from its return value. For example:

```
auto mul(int i, double d) { return i*d; }        // here, "auto" means "deduce the return type"
```

This can be convenient, especially for generic functions (§7.3.1) and lambdas (§7.3.3), but should be used carefully because a deduced type does not offer a stable interface: a change to the implementation of the function (or lambda) can change its type.

3.4.4 Suffix Return Type （ 返回类型后置）

Why does the return type come before a function's name and arguments? The reason is mostly historical. That's the way Fortran, C, and Simula did it (and still do). However, sometimes we need to look at the arguments to determine the type of the result. Return type deduction is one example of that, but not the only one. In examples beyond the scope of this book, the issue creeps up in connection with namespaces (§3.3), lambdas (§7.3.3), and concepts (§8.2). Consequently, we allow adding the return type after the argument list where we want to be explicit about the return type. That makes **auto** mean "the return type will be mentioned later or be deduced." For example:

```
auto mul(int i, double d) -> double { return i*d; }          // the return type is "double"
```

As with variables (§1.4.2), we can use this notation to line up names more neatly. Compare this use of suffix return types to the version in (§1.3):

```
auto next_elem() -> Elem*;
auto exit(int) -> void;
auto sqrt(double) -> double;
```

I find this suffix return type notation more logical than the traditional prefix one, but since the vast majority of code uses the traditional notation, I have stuck with that in this book.

3.4.5 Structured Binding（结构化绑定）

A function can return only a single value, but that value can be a class object with many members. This allows us to elegantly return many values. For example:

```
struct Entry {
      string name;
      int value;
};

Entry read_entry(istream& is)          // naive read function (for a better version, see §11.5)
{
      string s;
      int i;
      is >> s >> i;
      return {s,i};
}

auto e = read_entry(cin);

cout << "{ " << e.name << " , " << e.value << " }\n";
```

Here, {s,i} is used to construct the Entry return value. Similarly, we can "unpack" an Entry's members into local variables:

```
auto [n,v] = read_entry(is);
cout << "{ " << n << " , " << v << " }\n";
```

The auto [n,v] declares two local variables n and v with their types deduced from read_entry()'s return type. This mechanism for giving local names to members of a class object is called *structured binding*.

Consider another example:

```
map<string,int> m;
// ... fill m ...
for (const auto [key,value] : m)
      cout << "{" << key << "," << value << "}\n";
```

As usual, we can decorate auto with const and &. For example:

```
void incr(map<string,int>& m)        // increment the value of each element of m
{
    for (auto& [key,value] : m)
        ++value;
}
```

When structured binding is used for a class with no private data, it is easy to see how the binding is done: there must be the same number of names defined for the binding as there are data members in the class object, and each name introduced in the binding names the corresponding member. There will not be any difference in the object code quality compared to explicitly using a composite object. In particular, use of structured binding does not imply a copy of the **struct**. Furthermore, the return of a simple **struct** rarely involves a copy because simple return types can be constructed directly in the location where they are needed (§3.4.2). The use of structured binding is all about how best to express an idea.

It is also possible to handle classes where access is through member functions. For example:

```
complex<double> z = {1,2};
auto [re,im] = z+2;            // re=3; im=2
```

A **complex** has two data members, but its interface consists of access functions, such as **real()** and **imag()**. Mapping a **complex<double>** to two local variables, such as **re** and **im** is feasible and efficient, but the technique for doing so is beyond the scope of this book.

3.5 Advice（建议）

[1] Distinguish between declarations (used as interfaces) and definitions (used as implementations); §3.1.
[2] Prefer **modules** over headers (where **modules** are supported); §3.2.2.
[3] Use header files to represent interfaces and to emphasize logical structure; §3.2; [CG: SF.3].
[4] **#include** a header in the source file that implements its functions; §3.2; [CG: SF.5].
[5] Avoid non-inline function definitions in headers; §3.2; [CG: SF.2].
[6] Use namespaces to express logical structure; §3.3; [CG: SF.20].
[7] Use **using**-directives for transition, for foundational libraries (such as **std**), or within a local scope; §3.3; [CG: SF.6] [CG: SF.7].
[8] Don't put a **using**-directive in a header file; §3.3; [CG: SF.7].
[9] Pass "small" values by value and "large" values by reference; §3.4.1; [CG: F.16].
[10] Prefer pass-by-**const**-reference over plain pass-by-reference; §3.4.1; [CG: F.17].
[11] Return values as function-return values (rather than by out-parameters); §3.4.2; [CG: F.20] [CG: F.21].
[12] Don't overuse return-type deduction; §3.4.2.
[13] Don't overuse structured binding; a named return type often gives more readable code; §3.4.5.

4

Error Handling

（错误处理）

Don't interrupt me while I'm interrupting.
– Winston S. Churchill

- Introduction
- Exceptions
- Invariants
- Error-Handling Alternatives
- Assertions
  assert(); static_assert; noexcept
- Advice

4.1 Introduction（引言）

Error handling is a large and complex topic with concerns and ramifications that go far beyond language facilities into programming techniques and tools. However, C++ provides a few features to help. The major tool is the type system itself. Instead of painstakingly building up our applications from the built-in types (e.g., **char**, **int**, and **double**) and statements (e.g., **if**, **while**, and **for**), we build types (e.g., **string**, **map**, and **thread**) and algorithms (e.g., **sort()**, **find_if()**, and **draw_all()**) that are appropriate for our applications. Such higher-level constructs simplify our programming, limit our opportunities for mistakes (e.g., you are unlikely to try to apply a tree traversal to a dialog box), and increase the compiler's chances of catching errors. The majority of C++ language constructs are dedicated to the design and implementation of elegant and efficient abstractions (e.g., user-defined types and algorithms using them). One effect of using such abstractions is that the point where a run-time error can be detected is separated from the point where it can be handled. As programs grow, and especially when libraries are used extensively, standards for handling errors become important. It is a good idea to articulate a strategy for error handling early on in the development of a program.

4.2 Exceptions（异常）

Consider again the **Vector** example. What *ought* to be done when we try to access an element that is out of range for the vector from §2.3?

- The writer of **Vector** doesn't know what the user would like to have done in this case (the writer of **Vector** typically doesn't even know in which program the vector will be running).
- The user of **Vector** cannot consistently detect the problem (if the user could, the out-of-range access wouldn't happen in the first place).

Assuming that out-of-range access is a kind of error that we want to recover from, the solution is for the **Vector** implementer to detect the attempted out-of-range access and tell the user about it. The user can then take appropriate action. For example, **Vector::operator[]()** can detect an attempted out-of-range access and throw an **out_of_range** exception:

```
double& Vector::operator[](int i)
{
	if (!(0<i && i<size()))
		throw out_of_range{"Vector::operator[]"};
	return elem[i];
}
```

The **throw** transfers control to a handler for exceptions of type **out_of_range** in some function that directly or indirectly called **Vector::operator[]()**. To do that, the implementation will *unwind* the function call stack as needed to get back to the context of that caller. That is, the exception handling mechanism will exit scopes and functions as needed to get back to a caller that has expressed interest in handling that kind of exception, invoking destructors (§5.2.2) along the way as needed. For example:

```
void f(Vector& v)
{
	// ...
	try { // out_of_range exceptions thrown in this block are handled by the handler defined below
		compute1(v);			// might try to access beyond the end of v
		Vector v2 = compute2(v);// might try to access beyond the end of v
		compute3(v2);			// might try to access beyond the end of v2
	}
	catch (const out_of_range& err) {	// oops: out_of_range error
		// ... handle range error ...
		cerr << err.what() << '\n';
	}
	// ...
}
```

We put code for which we are interested in handling exceptions into a **try**-block. The calls of **compute1()**, **compute2()**, and **compute3()** are meant to represent code for which it is not simple to determine in advance if a range error will happen. The **catch**-clause is provided to handle exceptions of type **out_of_range**. Had **f()** not been a good place to handle such exceptions, we would not have used a **try**-block but instead let the exception implicitly pass to **f()**'s caller.

The **out_of_range** type is defined in the standard library (in **<stdexcept>**) and is in fact used by some standard-library container access functions.

I caught the exception by reference to avoid copying and used the what() function to print the error message put into it at the throw-point.

Use of the exception-handling mechanisms can make error handling simpler, more systematic, and more readable. To achieve that, don't overuse try-statements. In many programs there are typically dozens of function calls between a throw and a function that can reasonably handle the exception thrown. Thus, most functions should simply allow the exception to be propagated up the call stack. The main technique for making error handling simple and systematic (called *Resource Acquisition Is Initialization; RAII*) is explained in §5.2.2. The basic idea behind RAII is for a constructor to acquire the resources necessary for a class to operate and have the destructor release all resources, thus making resource release guaranteed and implicit.

4.3 Invariants（约束条件）

The use of exceptions to signal out-of-range access is an example of a function checking its argument and refusing to act because a basic assumption, a *precondition*, didn't hold. Had we formally specified Vector's subscript operator, we would have said something like "the index must be in the [0:size()) range," and that was in fact what we tested in our operator[](). The [a:b) notation specifies a half-open range, meaning that a is part of the range, but b is not. Whenever we define a function, we should consider what its preconditions are and consider whether to test them (§4.4). For most applications it is a good idea to test simple invariants; see also §4.5.

However, operator[]() operates on objects of type Vector and nothing it does makes any sense unless the members of Vector have "reasonable" values. In particular, we did say "elem points to an array of sz doubles" but we only said that in a comment. Such a statement of what is assumed to be true for a class is called a *class invariant*, or simply an *invariant*. It is the job of a constructor to establish the invariant for its class (so that the member functions can rely on it) and for the member functions to make sure that the invariant holds when they exit. Unfortunately, our Vector constructor only partially did its job. It properly initialized the Vector members, but it failed to check that the arguments passed to it made sense. Consider:

```
Vector v(−27);
```

This is likely to cause chaos.

Here is a more appropriate definition:

```
Vector::Vector(int s)
{
    if (s<0)
        throw length_error{"Vector constructor: negative size"};
    elem = new double[s];
    sz = s;
}
```

I use the standard-library exception length_error to report a negative number of elements because some standard-library operations use that exception to report problems of this kind. If operator new can't find memory to allocate, it throws a std::bad_alloc. We can now write:

```
void test(int n)
{
	try {
		Vector v(n);
	}
	catch (std::length_error& err) {
		// ... handle negative size ...
	}
	catch (std::bad_alloc& err) {
		// ... handle memory exhaustion ...
	}
}

void run()
{
	test(-27);		// throws length_error (-27 is too small)
	test(1'000'000'000);	// may throw bad_alloc
	test(10);		// likely OK
}
```

Memory exhaustion occurs if you ask for more memory than the machine offers or if your program already has consumed almost that much and your request pushes it over the limit. Note that modern operating systems typically will give you more space than will fit in physical memory at once, so asking for too much memory can cause serious slowdown long before triggering **bad_alloc**.

You can define your own classes to be used as exceptions and have them carry as little or as much information as you need from a point where an error is detected to a point where it can be handled (§4.2). It is not necessary to use the standard-library exception hierarchy.

Often, a function has no way of completing its assigned task after an exception is thrown. Then, "handling" an exception means doing some minimal local cleanup and rethrowing the exception. For example:

```
void test(int n)
{
	try {
		Vector v(n);
	}
	catch (std::length_error&) {	// do something and rethrow
		cerr << "test failed: length error\n";
		throw;	// rethrow
	}
	catch (std::bad_alloc&) {	// ouch! this program is not designed to handle memory exhaustion
		std::terminate();	// terminate the program
	}
}
```

In well-designed code **try**-blocks are rare. Avoid overuse by systematically using the RAII technique (§5.2.2, §6.3).

The notion of invariants is central to the design of classes, and preconditions serve a similar role in the design of functions:

- Formulating invariants helps us to understand precisely what we want.
- Invariants force us to be specific; this gives us a better chance of getting our code correct.

The notion of invariants underlies C++'s notions of resource management supported by constructors (Chapter 5) and destructors (§5.2.2, §15.2.1).

4.4 Error-Handling Alternatives（错误处理的其他替代方式）

Error handling is a major issue in all real-world software, so naturally there are a variety of approaches. If an error is detected and it cannot be handled locally in a function, the function must somehow communicate the problem to some caller. Throwing an exception is C++'s most general mechanism for that.

There are languages where exceptions are designed simply to provide an alternate mechanism for returning values. C++ is not such a language: exceptions are designed to be used to report failure to complete a given task. Exceptions are integrated with constructors and destructors to provide a coherent framework for error handling and resource management (§5.2.2, §6.3). Compilers are optimized to make returning a value much cheaper than throwing the same value as an exception.

Throwing an exception is not the only way of reporting an error that cannot be handled locally. A function can indicate that it cannot perform its allotted task by:

- throwing an exception
- somehow returning a value indicating failure
- terminating the program (by invoking a function like **terminate()**, **exit()**, or **abort()** (§16.8)).

We return an error indicator (an "error code") when:

- A failure is normal and expected. For example, it is quite normal for a request to open a file to fail (maybe there is no file of that name or maybe the file cannot be opened with the permissions requested).
- An immediate caller can reasonably be expected to handle the failure.
- An error happens in one of a set of parallel tasks and we need to know which task failed.
- A system has so little memory that the run-time support for exceptions would crowd out essential functionality.

We throw an exception when:

- An error is so rare that a programmer is likely to forget to check for it. For example, when did you last check the return value of **printf()**?
- An error cannot be handled by an immediate caller. Instead, the error has to percolate back up the call chain to an "ultimate caller." For example, it is infeasible to have every function in an application reliably handle every allocation failure and network outage. Repeatedly checking an error-code would be tedious, expensive, and error-prone. The tests for errors and passing error-codes as return values can easily obscure the main logic of a function.
- New kinds of errors can be added in lower-modules of an application so that higher-level modules are not written to cope with such errors. For example, when a previously single-threaded application is modified to use multiple threads or resources are placed remotely to be accessed over a network.
- No suitable return path for errors codes is available. For example, a constructor does not have a return value for a "caller" to check. In particular, constructors may be invoked for

several local variables or in a partially constructed complex object so that clean-up based on error codes would be quite complicated. Similarly, an operators don't usually have an obvious return path for error codes. For example, **a∗b+c/d**.
- The return path of a function is made more complicated or more expensive by a need to pass both a value and an error indicator back (e.g., a **pair**; §15.3.3), possibly leading to the use of out-parameters, non-local error-status indicators, or other workarounds.
- The recovery from errors depends on the results of several function calls, leading to the need to maintain local state between calls and complicated control structures.
- The function that found the error was a callback (a function argument), so the immediate caller may not even know what function was called.
- An error implies that some "undo action" is needed (§5.2.2).

We terminate when
- An error is of a kind from which we cannot recover. For example, for many – but not all – systems there is no reasonable way to recover from memory exhaustion.
- The system is one where error-handling is based on restarting a thread, process, or computer whenever a non-trivial error is detected.

One way to ensure termination is to add **noexcept** (§4.5.3) to a function so that a **throw** from anywhere in the function's implementation will turn into a **terminate()**. Note that there are applications that can't accept unconditional terminations, so alternatives must be used. A library for general-purpose use should never unconditionally terminate.

Unfortunately, these conditions are not always logically disjoint and easy to apply. The size and complexity of a program matters. Sometimes the tradeoffs change as an application evolves. Experience is required. When in doubt, prefer exceptions because their use scales better and doesn't require external tools to check that all errors are handled.

Don't believe that all error codes or all exceptions are bad; there are clear uses for both. Furthermore, do not believe the myth that exception handling is slow; it is often faster than correct handling of complex or rare error conditions, and of repeated tests of error codes.

RAII (§5.2.2, §6.3) is essential for simple and efficient error-handling using exceptions. Code littered with **try**-blocks often simply reflects the worst aspects of error-handling strategies conceived for error codes.

4.5 Assertions（断言）

There is currently no general and standard way of writing optional run-time tests of invariants, preconditions, etc. However, for many large programs, there is a need to support users who want to rely on extensive run-time checks while testing, but then deploy code with minimal checks.

For now, we have to rely on ad hoc mechanisms. There are many such mechanisms. They need to be flexible, general, and imply no cost when not enabled. This implies simplicity of conception and sophistication in implementation. Here is a scheme that I have used:

```
enum class Error_action { ignore, throwing, terminating, logging };          // error-handling alternatives

constexpr Error_action default_Error_action = Error_action::throwing;        // a default
```

```
enum class Error_code { range_error, length_error };                    // individual errors

string error_code_name[] { "range error", "length error" };             // names of individual errors

template<Error_action action = default_Error_action, class C>
constexpr void expect(C cond, Error_code x) // take "action" if the expected condition "cond" doesn't hold
{
  if constexpr (action == Error_action::logging)
    if (!cond()) std::cerr << "expect() failure: " << int(x) << ' ' << error_code_name[int(x)] << '\n';
  if constexpr (action == Error_action::throwing)
    if (!cond()) throw x;
  if constexpr (action == Error_action::terminating)
    if (!cond()) terminate();
  // or no action
}
```

This may seem mindboggling at first glance as many of the language features used are not yet presented. However, as required, it is both very flexible and trivial to use. For example:

```
double& Vector::operator[](int i)
{
      expect([i,this] { return 0<=i && i<size(); }, Error_code::range_error);
      return elem[i];
}
```

This checks if a subscript is in range and takes the default action, throwing an exception, if it is not. The condition expected to hold, **0<=i&&i<size()**, is passed to **expect()** as a lambda, **[i,this]{return 0<=i&&i<size();}** (§7.3.3). The **if constexpr** tests are done at compile time (§7.4.3) so at most one run-time test is performed for each call of **expect()**. Set **action** to **Error_action::ignore** and no action is taken and no code is generated for **expect()**.

By setting **default_Error_action** a user can select an action suitable for a particular deployment of the program, such as **terminating** or **logging**. To support logging, a table of **error_code_names** needs to be defined. The logging information could be improved by using **source_location** (§16.5).

In many systems, it is important that an assertion mechanism, such as **expect()**, offers a single point of control of the meaning of assertion failures. Searching a large code base for **if**-statements that are really checks of assumptions is typically impractical.

4.5.1 assert()

The standard library offers the debug macro, **assert()**, to assert that a condition must hold at run time. For example:

```
void f(const char* p)
{
      assert(p!=nullptr);   // p must not be the nullptr
      // ...
}
```

If the condition of an **assert()** fails in "debug mode," the program terminates. If not in debug mode, the **assert()** is not checked. That's pretty crude and inflexible, but often better than nothing.

4.5.2 Static Assertions（Static 断言）

Exceptions report errors found at run time. If an error can be found at compile time, it is usually preferable to do so. That's what much of the type system and the facilities for specifying the interfaces to user-defined types are for. However, we can also perform simple checks on most properties that are known at compile time and report failures to meet our expectations as compiler error messages. For example:

```
static_assert(4<=sizeof(int), "integers are too small");  // check integer size
```

This will write **integers are too small** if **4<=sizeof(int)** does not hold; that is, if an **int** on this system does not have at least 4 bytes. We call such statements of expectations *assertions*.

The **static_assert** mechanism can be used for anything that can be expressed in terms of constant expressions (§1.6). For example:

```
constexpr double C = 299792.458;                        // km/s

void f(double speed)
{
     constexpr double local_max = 160.0/(60*60);        // 160 km/h == 160.0/(60*60) km/s

     static_assert(speed<C,"can't go that fast");       // error: speed must be a constant
     static_assert(local_max<C,"can't go that fast");   // OK

     // ...
}
```

In general, **static_assert(A,S)** prints **S** as a compiler error message if **A** is not **true**. If you don't want a specific message printed, leave out the **S** and the compiler will supply a default message:

```
static_assert(4<=sizeof(int));       // use default message
```

The default message is typically the source location of the **static_assert** plus a character representation of the asserted predicate.

One important use of **static_assert** is to make assertions about types used as parameters in generic programming (§8.2, §16.4).

4.5.3 noexcept

A function that should never throw an exception can be declared **noexcept**. For example:

```
void user(int sz) noexcept
{
     Vector v(sz);
     iota(&v[0],&v[sz],1);      // fill v with 1,2,3,4... (see §17.3)
     // ...
}
```

If all good intent and planning fails, so that **user()** still throws, **std::terminate()** is called to immediately terminate the program.

Thoughtlessly sprinkling **noexcept** on functions is hazardous. If a **noexcept** function calls a function that throws an exception expecting it to be caught and handled, the **noexcept** turns that into a fatal error. Also, **noexcept** forces the writer to handle errors through some form of error codes that can be complex, error-prone, and expensive (§4.4). Like other powerful language features, **noexcept** should be applied with understanding and caution.

4.6 Advice（建议）

[1] Throw an exception to indicate that you cannot perform an assigned task; §4.4; [CG: E.2].
[2] Use exceptions for error handling only; §4.4; [CG: E.3].
[3] Failing to open a file or to reach the end of an iteration are expected events and not exceptional; §4.4.
[4] Use error codes when an immediate caller is expected to handle the error; §4.4.
[5] Throw an exception for errors expected to percolate up through many function calls; §4.4.
[6] If in doubt whether to use an exception or an error code, prefer exceptions; §4.4.
[7] Develop an error-handling strategy early in a design; §4.4; [CG: E.12].
[8] Use purpose-designed user-defined types as exceptions (not built-in types); §4.2.
[9] Don't try to catch every exception in every function; §4.4; [CG: E.7].
[10] You don't have to use the standard-library exception class hierarchy; §4.3.
[11] Prefer RAII to explicit **try**-blocks; §4.2, §4.3; [CG: E.6].
[12] Let a constructor establish an invariant, and throw if it cannot; §4.3; [CG: E.5].
[13] Design your error-handling strategy around invariants; §4.3; [CG: E.4].
[14] What can be checked at compile time is usually best checked at compile time; §4.5.2 [CG: P.4] [CG: P.5].
[15] Use an assertion mechanism to provide a single point of control of the meaning of failure; §4.5.
[16] Concepts (§8.2) are compile-time predicates and therefore often useful in assertions; §4.5.2.
[17] If your function may not throw, declare it **noexcept**; §4.4; [CG: E.12].
[18] Don't apply **noexcept** thoughtlessly; §4.5.3.

5

Classes
(类)

Those types are not "abstract";
they are as real as int *and* float.
– Doug McIlroy

5.1 Introduction（引言）

This chapter and the next three aim to give you an idea of C++'s support for abstraction and resource management without going into a lot of detail:

- This chapter informally presents ways of defining and using new types (*user-defined types*). In particular, it presents the basic properties, implementation techniques, and language facilities used for *concrete classes*, *abstract classes*, and *class hierarchies*.
- Chapter 6 presents the operations that have defined meaning in C++, such as constructors, destructors, and assignments. It outlines the rules for using those in combination to control the life cycle of objects and to support simple, efficient, and complete resource management.
- Chapter 7 introduces templates as a mechanism for parameterizing types and algorithms with other types and algorithms. Computations on user-defined and built-in types are represented as functions, sometimes generalized to *function templates* and *function objects*.

- Chapter 8 gives an overview of the concepts, techniques, and language features that underlie generic programming. The focus is on the definition and use of *concepts* for precisely specifying interfaces to templates and guide design. *Variadic templates* are introduced for specifying the most general and most flexible interfaces.

These are the language facilities supporting the programming styles known as *object-oriented programming* and *generic programming*. Chapters 9–18 follow up by presenting examples of standard-library facilities and their use.

5.1.1 Classes（类的概述）

The central language feature of C++ is the *class*. A class is a user-defined type provided to represent an entity in the code of a program. Whenever our design for a program has a useful idea, entity, collection of data, etc., we try to represent it as a class in the program so that the idea is there in the code, rather than just in our heads, in a design document, or in some comments. A program built out of a well-chosen set of classes is far easier to understand and get right than one that builds everything directly in terms of the built-in types. In particular, classes are often what libraries offer.

Essentially all language facilities beyond the fundamental types, operators, and statements exist to help define better classes or to use them more conveniently. By "better," I mean more correct, easier to maintain, more efficient, more elegant, easier to use, easier to read, and easier to reason about. Most programming techniques rely on the design and implementation of specific kinds of classes. The needs and tastes of programmers vary immensely. Consequently, the support for classes is extensive. Here, we consider the basic support for three important kinds of classes:

- Concrete classes (§5.2)
- Abstract classes (§5.3)
- Classes in class hierarchies (§5.5)

An astounding number of useful classes turn out to be of one of these three kinds. Even more classes can be seen as simple variants of these kinds or are implemented using combinations of the techniques used for these.

5.2 Concrete Types（具体类型）

The basic idea of *concrete classes* is that they behave "just like built-in types." For example, a complex number type and an infinite-precision integer are much like a built-in `int`, except of course that they have their own semantics and sets of operations. Similarly, a `vector` and a `string` are much like built-in arrays, except that they are more flexible and better behaved (§10.2, §11.3, §12.2).

The defining characteristic of a concrete type is that its representation is part of its definition. In many important cases, such as a `vector`, that representation is only one or more pointers to data stored elsewhere, but that representation is present in each object of the concrete class. That allows implementations to be optimally efficient in time and space. In particular, it allows us to

- Place objects of concrete types on the stack, in statically allocated memory, and in other objects (§1.5).
- Refer to objects directly (and not just through pointers or references).
- Initialize objects immediately and completely (e.g., using constructors; §2.3).

- Copy and move objects (§6.2).

The representation can be private and accessible only through the member functions (as it is for Vector; §2.3), but it is present. Therefore, if the representation changes in any significant way, a user must recompile. This is the price to pay for having concrete types behave exactly like built-in types. For types that don't change often, and where local variables provide much-needed clarity and efficiency, this is acceptable and often ideal. To increase flexibility, a concrete type can keep major parts of its representation on the free store (dynamic memory, heap) and access them through the part stored in the class object itself. That's the way vector and string are implemented; they can be considered resource handles with carefully crafted interfaces.

5.2.1 An Arithmetic Type（一种算术类型）

The "classical user-defined arithmetic type" is complex:

```
class complex {
      double re, im;  // representation: two doubles
public:
      complex(double r, double i) :re{r}, im{i} {}         // construct complex from two scalars
      complex(double r) :re{r}, im{0} {}                   // construct complex from one scalar
      complex() :re{0}, im{0} {}                           // default complex: {0,0}
      complex(complex z) :re{z.re}, im{z.im} {}            // copy constructor

      double real() const { return re; }
      void real(double d) { re=d; }
      double imag() const { return im; }
      void imag(double d) { im=d; }

      complex& operator+=(complex z)
      {
            re+=z.re;              // add to re and im
            im+=z.im;
            return *this;          // return the result
      }

      complex& operator-=(complex z)
      {
            re-=z.re;
            im-=z.im;
            return *this;
      }

      complex& operator*=(complex);    // defined out-of-class somewhere
      complex& operator/=(complex);    // defined out-of-class somewhere
};
```

This is a simplified version of the standard-library complex (§17.4). The class definition itself contains only the operations requiring access to the representation. The representation is simple and conventional. For practical reasons, it has to be compatible with what Fortran provided 60 years ago, and we need a conventional set of operators. In addition to the logical demands, complex must be efficient or it will remain unused. This implies that simple operations must be inlined. That is, simple operations (such as constructors, +=, and **imag()**) must be implemented without function calls in the generated machine code. Functions defined in a class are inlined by default. It is possible to explicitly request inlining by preceding a function declaration with the keyword **inline**. An industrial-strength **complex** (like the standard-library one) is carefully implemented to do appropriate inlining. In addition, the standard-library **complex** has the functions shown here declared **constexpr** so that we can do complex arithmetic at compile time.

Copy assignment and copy initialization are implicitly defined (§6.2).

A constructor that can be invoked without an argument is called a *default constructor*. Thus, **complex()** is **complex**'s default constructor. By defining a default constructor you eliminate the possibility of uninitialized variables of that type.

The **const** specifiers on the functions returning the real and imaginary parts indicate that these functions do not modify the object for which they are called. A **const** member function can be invoked for both **const** and non-**const** objects, but a non-**const** member function can only be invoked for non-**const** objects. For example:

```
complex z = {1,0};
const complex cz {1,3};
z = cz;                 // OK: assigning to a non-const variable
cz = z;                 // error: assignment to a const
double x = z.real();    // OK: complex::real() is const
```

Many useful operations do not require direct access to the representation of **complex**, so they can be defined separately from the class definition:

```
complex operator+(complex a, complex b) { return a+=b; }
complex operator-(complex a, complex b) { return a-=b; }
complex operator-(complex a) { return {-a.real(), -a.imag()}; }     // unary minus
complex operator*(complex a, complex b) { return a*=b; }
complex operator/(complex a, complex b) { return a/=b; }
```

Here, I use the fact that an argument passed by value is copied so that I can modify an argument without affecting the caller's copy and use the result as the return value.

The definitions of == and != are straightforward:

```
bool operator==(complex a, complex b) { return a.real()==b.real() && a.imag()==b.imag(); }  // equal
bool operator!=(complex a, complex b) { return !(a==b); }                                   // not equal
```

Class **complex** can be used like this:

```
void f(complex z)
{
        complex a {2.3};            // construct {2.3,0.0} from 2.3
        complex b {1/a};
        complex c {a+z*complex{1,2.3}};
        if (c != b)
                c = -(b/a)+2*b;
}
```

The compiler converts operators involving **complex** numbers into appropriate function calls. For example, **c!=b** means **operator!=(c,b)** and **1/a** means **operator/(complex{1},a)**.

User-defined operators (“overloaded operators”) should be used cautiously and conventionally (§6.4). The syntax is fixed by the language, so you can’t define a unary /. Also, it is not possible to change the meaning of an operator for built-in types, so you can’t redefine + to subtract ints.

5.2.2 A Container（容器）

A *container* is an object holding a collection of elements. We call class Vector a container because objects of type Vector are containers. As defined in §2.3, Vector isn’t an unreasonable container of doubles: it is simple to understand, establishes a useful invariant (§4.3), provides range-checked access (§4.2), and provides size() to allow us to iterate over its elements. However, it does have a fatal flaw: it allocates elements using new but never deallocates them. That’s not a good idea because C++ does not offer a garbage collector to make unused memory available for new objects. In some environments you can’t use a collector, and often you prefer more precise control of destruction for logical or performance reasons. We need a mechanism to ensure that the memory allocated by the constructor is deallocated; that mechanism is a *destructor*:

```
class Vector {
public:
     Vector(int s) :elem{new double[s]}, sz{s}          // constructor: acquire resources
     {
          for (int i=0; i!=s; ++i)               // initialize elements
               elem[i]=0;
     }

     ~Vector() { delete[] elem; }                       // destructor: release resources

     double& operator[](int i);
     int size() const;
private:
     double* elem;          // elem points to an array of sz doubles
     int sz;
};
```

The name of a destructor is the complement operator, ~, followed by the name of the class; it is the complement of a constructor.

Vector’s constructor allocates some memory on the free store (also called the *heap* or *dynamic memory*) using the new operator. The destructor cleans up by freeing that memory using the delete[] operator. Plain delete deletes an individual object; delete[] deletes an array.

This is all done without intervention by users of Vector. The users simply create and use Vectors much as they would variables of built-in types. For example:

```
Vector gv(10);                          // global variable; gv is destroyed at the end of the program

Vector* gp = new Vector(100); // Vector on free store; never implicitly destroyed
```

```
void fct(int n)
{
	Vector v(n);
	// ... use v ...
	{
		Vector v2(2*n);
		// ... use v and v2 ...
	} // v2 is destroyed here
	// ... use v ..
} // v is destroyed here
```

Vector obeys the same rules for naming, scope, allocation, lifetime, etc. (§1.5), as does a built-in type, such as int and char. This Vector has been simplified by leaving out error handling; see §4.4.

The constructor/destructor combination is the basis of many elegant techniques. In particular, it is the basis for most C++ general resource management techniques (§6.3, §15.2.1). Consider a graphical illustration of a Vector:

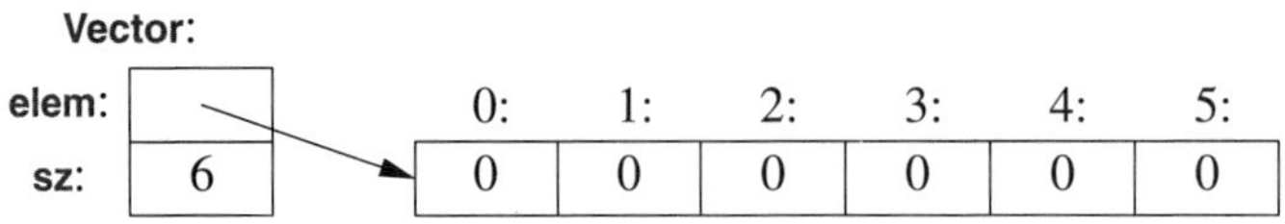

Vector:

			0:	1:	2:	3:	4:	5:
elem:	(pointer)	→						
sz:	6		0	0	0	0	0	0

The constructor allocates the elements and initializes the Vector members appropriately. The destructor deallocates the elements. This *handle-to-data model* is very commonly used to manage data that can vary in size during the lifetime of an object. The technique of acquiring resources in a constructor and releasing them in a destructor, known as *Resource Acquisition Is Initialization* or *RAII*. It allows us to eliminate "naked new operations," that is, to avoid allocations in general code and keep them buried inside the implementation of well-behaved abstractions. Similarly, "naked delete operations" should be avoided. Avoiding naked new and naked delete makes code far less error-prone and far easier to keep free of resource leaks (§15.2.1).

5.2.3 Initializing Containers（容器的初始化）

A container exists to hold elements, so obviously we need convenient ways of getting elements into a container. We can create a Vector with an appropriate number of elements and then assign to those later, but typically other ways are more elegant. Here, I just mention two favorites:

- *Initializer-list constructor*: Initialize with a list of elements.
- push_back(): Add a new element at the end of (at the back of) the sequence.

These can be declared like this:

```
class Vector {
public:
	Vector();                                   // default initalize to "empty"; that is, to no elements
	Vector(std::initializer_list<double>);      // initialize with a list of doubles
	// ...
	void push_back(double);                     // add element at end, increasing the size by one
	// ...
};
```

The push_back() is useful for input of arbitrary numbers of elements. For example:

```
Vector read(istream& is)
{
      Vector v;
      for (double d; is>>d; )          // read floating-point values into d
            v.push_back(d);            // add d to v
      return v;
}
```

The input loop is terminated by an end-of-file or a formatting error. Until that happens, each number read is added to the Vector so that at the end, v's size is the number of elements read. I used a for-statement rather than the more conventional while-statement to limit the scope of d to the loop.

Returning a potentially huge amount of data from read() could be expensive. The way to guarantee that returning a Vector is cheap is to provide it with a move constructor (§6.2.2):

```
Vector v = read(cin);      // no copy of Vector elements here
```

The way that std::vector is represented to make push_back() and other operations that change a vector's size efficient is presented in §12.2.

The std::initializer_list used to define the initializer-list constructor is a standard-library type known to the compiler: when we use a {}-list, such as {1,2,3,4}, the compiler will create an object of type initializer_list to give to the program. So, we can write:

```
Vector v1 = {1, 2, 3, 4, 5};             // v1 has 5 elements
Vector v2 = {1.23, 3.45, 6.7, 8};        // v2 has 4 elements
```

Vector's initializer-list constructor might be defined like this:

```
Vector::Vector(std::initializer_list<double> lst)      // initialize with a list
      :elem{new double[lst.size()]}, sz{static_cast<int>(lst.size())}
{
      copy(lst.begin(),lst.end(),elem);               // copy from lst into elem (§13.5)
}
```

Unfortunately, the standard-library uses unsigned integers for sizes and subscripts, so we need to use the ugly static_cast to explicitly convert the size of the initializer list to an int. This is pedantic because the chance that the number of elements in a handwritten list is larger than the largest integer (32,767 for 16-bit integers and 2,147,483,647 for 32-bit integers) is rather low. However, the type system has no common sense. It knows about the possible values of variables, rather than actual values, so it might complain where there is no actual violation. Such warnings can occasionally save the programmer from a bad error.

A static_cast does not check the value it is converting; the programmer is trusted to use it correctly. This is not always a good assumption, so if in doubt, check the value. Explicit type conversions (often called *casts* to remind you that they are used to prop up something broken) are best avoided. Try to use unchecked casts only for the lowest level of a system. They are error-prone.

Other casts are reinterpret_cast and bit_cast (§16.7) for treating an object as simply a sequence of bytes and const_cast for "casting away const." Judicious use of the type system and well-designed libraries allow us to eliminate unchecked casts in higher-level software.

5.3 Abstract Types（抽象类型）

Types such as complex and Vector are called *concrete types* because their representation is part of their definition. In that, they resemble built-in types. In contrast, an *abstract type* is a type that completely insulates a user from implementation details. To do that, we decouple the interface from the representation and give up genuine local variables. Since we don't know anything about the representation of an abstract type (not even its size), we must allocate objects on the free store (§5.2.2) and access them through references or pointers (§1.7, §15.2.1).

First, we define the interface of a class Container, which we will design as a more abstract version of our Vector:

```
class Container {
public:
        virtual double& operator[](int) = 0;        // pure virtual function
        virtual int size() const = 0;               // const member function (§5.2.1)
        virtual ~Container() {}                     // destructor (§5.2.2)
};
```

This class is a pure interface to specific containers defined later. The word virtual means "may be redefined later in a class derived from this one." Unsurprisingly, a function declared virtual is called a *virtual function*. A class derived from Container provides an implementation for the Container interface. The curious =0 syntax says the function is *pure virtual*; that is, some class derived from Container *must* define the function. Thus, it is not possible to define an object that is just a Container. For example:

```
Container c;                                // error: there can be no objects of an abstract class
Container* p = new Vector_container(10);    // OK: Container is an interface for Vector_container
```

A Container can only serve as the interface to a class that implements its operator[]() and size() functions. A class with a pure virtual function is called an *abstract class*.

This Container can be used like this:

```
void use(Container& c)
{
        const int sz = c.size();

        for (int i=0; i!=sz; ++i)
                cout << c[i] << '\n';
}
```

Note how use() uses the Container interface in complete ignorance of implementation details. It uses size() and [ ] without any idea of exactly which type provides their implementation. A class that provides the interface to a variety of other classes is often called a *polymorphic type*.

As is common for abstract classes, Container does not have a constructor. After all, it does not have any data to initialize. On the other hand, Container does have a destructor and that destructor is virtual, so that classes derived from Container can provide implementations. Again, that is common for abstract classes because they tend to be manipulated through references or pointers, and someone destroying a Container through a pointer has no idea what resources are owned by its implementation; see also §5.5.

The abstract class **Container** defines only an interface and no implementation. For **Container** to be useful, we have to implement a container that implements the functions required by its interface. For that, we could use the concrete class **Vector**:

```
class Vector_container : public Container {   // Vector_container implements Container
public:
      Vector_container(int s) : v(s) { }      // Vector of s elements
      ~Vector_container() {}

      double& operator[](int i) override { return v[i]; }
      int size() const override { return v.size(); }
private:
      Vector v;
};
```

The **:public** can be read as "is derived from" or "is a subtype of." Class **Vector_container** is said to be *derived* from class **Container**, and class **Container** is said to be a *base* of class **Vector_container**. An alternative terminology calls **Vector_container** and **Container** *subclass* and *superclass*, respectively. The derived class is said to inherit members from its base class, so the use of base and derived classes is commonly referred to as *inheritance*.

The members **operator[]()** and **size()** are said to *override* the corresponding members in the base class **Container**. I used the explicit **override** to make clear what's intended. The use of **override** is optional, but being explicit allows the compiler to catch mistakes, such as misspellings of function names or slight differences between the type of a **virtual** function and its intended overrider. The explicit use of **override** is particularly useful in larger class hierarchies where it can otherwise be hard to know what is supposed to override what.

The destructor (**~Vector_container()**) overrides the base class destructor (**~Container()**). Note that the member destructor (**~Vector()**) is implicitly invoked by its class's destructor (**~Vector_container()**).

For a function like **use(Container&)** to use a **Container** in complete ignorance of implementation details, some other function will have to make an object on which it can operate. For example:

```
void g()
{
      Vector_container vc(10);        // Vector of ten elements
      // ... fill vc ...
      use(vc);
}
```

Since **use()** doesn't know about **Vector_containers** but only knows the **Container** interface, it will work just as well for a different implementation of a **Container**. For example:

```
class List_container : public Container {     // List_container implements Container
public:
      List_container() { }        // empty List
      List_container(initializer_list<double> il) : ld{il} { }
      ~List_container() {}

      double& operator[](int i) override;
      int size() const override { return ld.size(); }
```

```
private:
      std::list<double> ld;        // (standard-library) list of doubles (§12.3)
};

double& List_container::operator[](int i)
{
      for (auto& x : ld) {
            if (i==0)
                  return x;
            --i;
      }
      throw out_of_range{"List container"};
}
```

Here, the representation is a standard-library **list<double>**. Usually, I would not implement a container with a subscript operation using a **list**, because performance of **list** subscripting is atrocious compared to **vector** subscripting. However, here I just wanted to show an implementation that is radically different from the usual one.

A function can create a **List_container** and have **use()** use it:

```
void h()
{
      List_container lc = {1, 2, 3, 4, 5, 6, 7, 8, 9};
      use(lc);
}
```

The point is that **use(Container&)** has no idea if its argument is a **Vector_container**, a **List_container**, or some other kind of container; it doesn't need to know. It can use any kind of **Container**. It knows only the interface defined by **Container**. Consequently, **use(Container&)** needn't be recompiled if the implementation of **List_container** changes or a brand-new class derived from **Container** is used.

The flip side of this flexibility is that objects must be manipulated through pointers or references (§6.2, §15.2.1).

5.4 Virtual Functions（虚函数）

Consider again the use of **Container**:

```
void use(Container& c)
{
      const int sz = c.size();

      for (int i=0; i!=sz; ++i)
            cout << c[i] << '\n';
}
```

How is the call **c[i]** in **use()** resolved to the right **operator[]()**? When **h()** calls **use()**, **List_container**'s **operator[]()** must be called. When **g()** calls **use()**, **Vector_container**'s **operator[]()** must be called. To achieve this resolution, a **Container** object must contain information to allow it to select the right function to call at run time. The usual implementation technique is for the compiler to convert the name of a virtual function into an index into a table of pointers to functions. That table is usually called the *virtual function table* or simply the **vtbl**. Each class with virtual functions has its own **vtbl** identifying its virtual functions. This can be represented graphically like this:

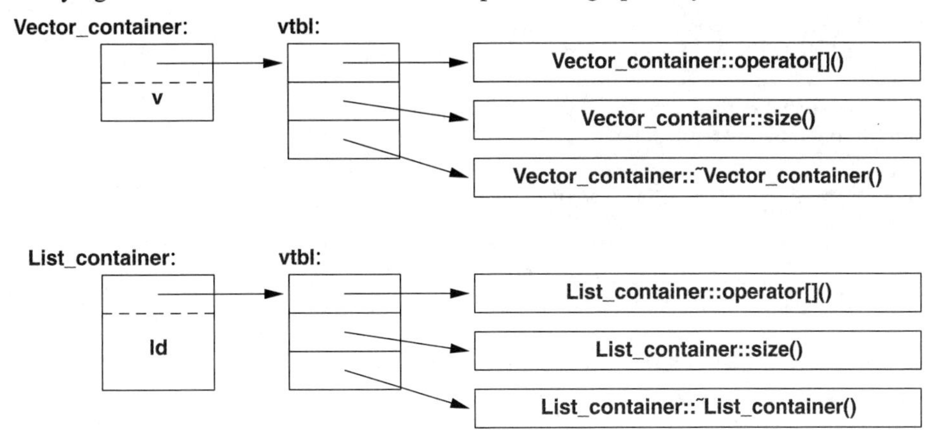

The functions in the **vtbl** allow the object to be used correctly even when the size of the object and the layout of its data are unknown to the caller. The implementation of the caller needs only to know the location of the pointer to the **vtbl** in a **Container** and the index used for each virtual function. This virtual call mechanism can be made almost as efficient as the "normal function call" mechanism (within 25% and far cheaper for repeated calls to the same object). Its space overhead is one pointer in each object of a class with virtual functions plus one **vtbl** for each such class.

5.5 Class Hierarchies（类层次结构）

The **Container** example is a very simple example of a class hierarchy. A *class hierarchy* is a set of classes ordered in a lattice created by derivation (e.g., **: public**). We use class hierarchies to represent concepts that have hierarchical relationships, such as "A fire engine is a kind of a truck that is a kind of a vehicle" and "A smiley face is a kind of a circle that is a kind of a shape." Huge hierarchies, with hundreds of classes, that are both deep and wide are common. As a semi-realistic classic example, let's consider shapes on a screen:

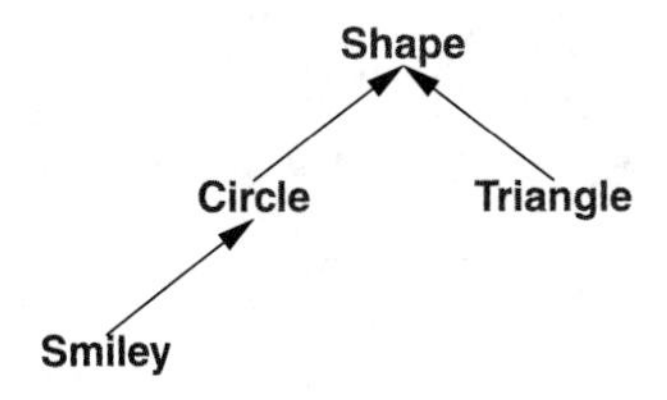

The arrows represent inheritance relationships. For example, class **Circle** is derived from class **Shape**. A class hierarchy is conventionally drawn growing down from the most basic class, the root, towards the (later defined) derived classes. To represent that simple diagram in code, we must first specify a class that defines the general properties of all shapes:

```
class Shape {
public:
	virtual Point center() const =0;		// pure virtual
	virtual void move(Point to) =0;

	virtual void draw() const = 0;			// draw on current "Canvas"
	virtual void rotate(int angle) = 0;

	virtual ~Shape() {}						// destructor
	// ...
};
```

Naturally, this interface is an abstract class: as far as representation is concerned, *nothing* (except the location of the pointer to the **vtbl**) is common for every **Shape**. Given this definition, we can write general functions manipulating vectors of pointers to shapes:

```
void rotate_all(vector<Shape*>& v, int angle) // rotate v's elements by angle degrees
{
	for (auto p : v)
		p->rotate(angle);
}
```

To define a particular shape, we must say that it is a **Shape** and specify its particular properties (including its virtual functions):

```
class Circle : public Shape {
public:
	Circle(Point p, int rad) :x{p}. r{rad} {}		// constructor

	Point center() const override { return x; }
	void move(Point to) override { x = to; }
	void draw() const override;
	void rotate(int) override {}			// nice simple algorithm
private:
	Point x;	// center
	int r;		// radius
};
```

So far, the **Shape** and **Circle** example provides nothing new compared to the **Container** and **Vector_container** example, but we can build further:

```
class Smiley : public Circle {	// use the circle as the base for a face
public:
	Smiley(Point p, int rad) : Circle{p,rad}, mouth{nullptr} { }
	~Smiley()
	{
		delete mouth;
		for (auto p : eyes)
			delete p;
	}
```

```
        void move(Point to) override;

        void draw() const override;
        void rotate(int) override;

        void add_eye(Shape* s)
        {
                eyes.push_back(s);
        }

        void set_mouth(Shape* s);
        virtual void wink(int i);           // wink eye number i

        // ...

private:
        vector<Shape*> eyes;                // usually two eyes
        Shape* mouth;
};
```

The **push_back()** member of **vector** copies its argument into the **vector** (here, **eyes**) as the last element, increasing that vector's size by one.

We can now define **Smiley::draw()** using calls to **Smiley**'s base and member **draw()**s:

```
void Smiley::draw() const
{
        Circle::draw();
        for (auto p : eyes)
                p->draw();
        mouth->draw();
}
```

Note the way that **Smiley** keeps its eyes in a standard-library **vector** and deletes them in its destructor. **Shape**'s destructor is **virtual** and **Smiley**'s destructor overrides it. A **virtual** destructor is essential for an abstract class because an object of a derived class is usually manipulated through the interface provided by its abstract base class. In particular, it may be **delete**d through a pointer to a base class. Then, the virtual function call mechanism ensures that the proper destructor is called. That destructor then implicitly invokes the destructors of its bases and members.

In this simplified example, it is the programmer's task to place the eyes and mouth appropriately within the circle representing the face.

We can add data members, operations, or both as we define a new class by derivation. This gives great flexibility with corresponding opportunities for confusion and poor design.

5.5.1 Benefits from Hierarchies（类层次结构的益处）

A class hierarchy offers two kinds of benefits:

- *Interface inheritance*: An object of a derived class can be used wherever an object of a base class is required. That is, the base class acts as an interface for the derived class. The **Container** and **Shape** classes are examples. Such classes are often abstract classes.

- *Implementation inheritance*: A base class provides functions or data that simplifies the implementation of derived classes. **Smiley**'s uses of **Circle**'s constructor and of **Circle::draw()** are examples. Such base classes often have data members and constructors.

Concrete classes – especially classes with small representations – are much like built-in types: we define them as local variables, access them using their names, copy them around, etc. Classes in class hierarchies are different: we tend to allocate them on the free store using **new**, and we access them through pointers or references. For example, consider a function that reads data describing shapes from an input stream and constructs the appropriate **Shape** objects:

```
enum class Kind { circle, triangle, smiley };

Shape* read_shape(istream& is)    // read shape descriptions from input stream is
{
     // ... read shape header from is and find its Kind k ...

     switch (k) {
     case Kind::circle:
          // ... read circle data {Point,int} into p and r ...
          return new Circle{p,r};
     case Kind::triangle:
          // ... read triangle data {Point,Point,Point} into p1, p2, and p3 ...
          return new Triangle{p1,p2,p3};
     case Kind::smiley:
          // ... read smiley data {Point,int,Shape,Shape,Shape} into p, r, e1, e2, and m ...
          Smiley* ps = new Smiley{p,r};
          ps->add_eye(e1);
          ps->add_eye(e2);
          ps->set_mouth(m);
          return ps;
     }
}
```

A program may use that shape reader like this:

```
void user()
{
     std::vector<Shape*> v;

     while (cin)
          v.push_back(read_shape(cin));

     draw_all(v);                  // call draw() for each element
     rotate_all(v,45);             // call rotate(45) for each element

     for (auto p : v)              // remember to delete elements
          delete p;
}
```

Obviously, the example is simplified – especially with respect to error handling – but it vividly illustrates that **user()** has absolutely no idea of which kinds of shapes it manipulates. The **user()**

code can be compiled once and later used for new **Shapes** added to the program. Note that there are no pointers to the shapes outside **user()**, so **user()** is responsible for deallocating them. This is done with the **delete** operator and relies critically on **Shape**'s virtual destructor. Because that destructor is virtual, **delete** invokes the destructor for the most derived class. This is crucial because a derived class may have acquired all kinds of resources (such as file handles, locks, and output streams) that need to be released. In this case, a **Smiley** deletes its **eyes** and **mouth** objects. Once it has done that, it calls **Circle**'s destructor. Objects are constructed "bottom up" (base first) by constructors and destroyed "top down" (derived first) by destructors.

5.5.2 Hierarchy Navigation （类层次结构导航）

The **read_shape()** function returns a **Shape*** so that we can treat all **Shapes** alike. However, what can we do if we want to use a member function that is only provided by a particular derived class, such as **Smiley**'s **wink()**? We can ask "is this **Shape** a kind of **Smiley**?" using the **dynamic_cast** operator:

```
Shape* ps {read_shape(cin)};

if (Smiley* p = dynamic_cast<Smiley*>(ps)) { // does ps point to a Smiley?
      // ... a Smiley; use it ...
}
else {
      // ... not a Smiley, try something else ...
}
```

If at run time the object pointed to by the argument of **dynamic_cast** (here, **ps**) is not of the expected type (here, **Smiley**) or a class derived from the expected type, **dynamic_cast** returns **nullptr**.

We use **dynamic_cast** to a pointer type when a pointer to an object of a different derived class is a valid argument. We then test whether the result is **nullptr**. This test can often conveniently be placed in the initialization of a variable in a condition.

When a different type is unacceptable, we can simply **dynamic_cast** to a reference type. If the object is not of the expected type, **dynamic_cast** throws a **bad_cast** exception:

```
Shape* ps {read_shape(cin)};
Smiley& r {dynamic_cast<Smiley&>(*ps)};      // somewhere, catch std::bad_cast
```

Code is cleaner when **dynamic_cast** is used with restraint. If we can avoid testing type information at run time, we can write simpler and more efficient code, but occasionally type information is lost and must be recovered. This typically happens when we pass an object to some system that accepts an interface specified by a base class. When that system later passes the object back to us, we might have to recover the original type. Operations similar to **dynamic_cast** are known as "is kind of" and "is instance of" operations.

5.5.3 Avoiding Resource Leaks （避免资源泄漏）

A *leak* is the conventional term for the what happens when we acquire a resource and fail to release it. Leaking resources must be avoided because a leak makes the leaked resource unavailable to the system. Thus, leaks can eventually lead to slowdown or even crashes as a system runs out of needed resources.

Experienced programmers will have noticed that I left open three opportunities for mistakes in the **Smiley** example:

- The implementer of **Smiley** may fail to **delete** the pointer to **mouth**.
- A user of **read_shape()** might fail to **delete** the pointer returned.
- The owner of a container of **Shape** pointers might fail to **delete** the objects pointed to.

In that sense, pointers to objects allocated on the free store are dangerous: a "plain old pointer" should not be used to represent ownership. For example:

```
void user(int x)
{
	Shape* p = new Circle{Point{0,0},10};
	// ...
	if (x<0) throw Bad_x{};	// potential leak
	if (x==0) return;		// potential leak
	// ...
	delete p;
}
```

This will leak unless **x** is positive. Assigning the result of **new** to a "naked pointer" is asking for trouble.

One simple solution to such problems is to use a standard-library **unique_ptr** (§15.2.1) rather than a "naked pointer" when deletion is required:

```
class Smiley : public Circle {
	// ...
private:
	vector<unique_ptr<Shape>> eyes; // usually two eyes
	unique_ptr<Shape> mouth;
};
```

This is an example of a simple, general, and efficient technique for resource management (§6.3).

As a pleasant side effect of this change, we no longer need to define a destructor for **Smiley**. The compiler will implicitly generate one that does the required destruction of the **unique_ptrs** (§6.3) in the **vector**. The code using **unique_ptr** will be exactly as efficient as code using the raw pointers correctly.

Now consider users of **read_shape()**:

```
unique_ptr<Shape> read_shape(istream& is) // read shape descriptions from input stream is
{
	// ... read shape header from is and find its Kind k ...

	switch (k) {
	case Kind::circle:
		// ... read circle data {Point,int} into p and r ...
		return unique_ptr<Shape>{new Circle{p,r}};		// §15.2.1
	// ...
}
```

```
void user()
{
    vector<unique_ptr<Shape>> v;

    while (cin)
        v.push_back(read_shape(cin));

    draw_all(v);                  // call draw() for each element
    rotate_all(v,45);             // call rotate(45) for each element
} // all Shapes implicitly destroyed
```

Now each object is owned by a **unique_ptr** that will **delete** the object when it is no longer needed, that is, when its **unique_ptr** goes out of scope.

For the **unique_ptr** version of **user()** to work, we need versions of **draw_all()** and **rotate_all()** that accept **vector<unique_ptr<Shape>>**s. Writing many such **_all()** functions could become tedious, so §7.3.2 shows an alternative.

5.6 Advice（建议）

[1] Express ideas directly in code; §5.1; [CG: P.1].

[2] A concrete type is the simplest kind of class. Where applicable, prefer a concrete type over more complicated classes and over plain data structures; §5.2; [CG: C.10].

[3] Use concrete classes to represent simple concepts; §5.2.

[4] Prefer concrete classes over class hierarchies for performance-critical components; §5.2.

[5] Define constructors to handle initialization of objects; §5.2.1, §6.1.1; [CG: C.40] [CG: C.41].

[6] Make a function a member only if it needs direct access to the representation of a class; §5.2.1; [CG: C.4].

[7] Define operators primarily to mimic conventional usage; §5.2.1; [CG: C.160].

[8] Use nonmember functions for symmetric operators; §5.2.1; [CG: C.161].

[9] Declare a member function that does not modify the state of its object **const**; §5.2.1.

[10] If a constructor acquires a resource, its class needs a destructor to release the resource; §5.2.2; [CG: C.20].

[11] Avoid “naked” **new** and **delete** operations; §5.2.2; [CG: R.11].

[12] Use resource handles and RAII to manage resources; §5.2.2; [CG: R.1].

[13] If a class is a container, give it an initializer-list constructor; §5.2.3; [CG: C.103].

[14] Use abstract classes as interfaces when complete separation of interface and implementation is needed; §5.3; [CG: C.122].

[15] Access polymorphic objects through pointers and references; §5.3.

[16] An abstract class typically doesn’t need a constructor; §5.3; [CG: C.126].

[17] Use class hierarchies to represent concepts with inherent hierarchical structure; §5.5.

[18] A class with a virtual function should have a virtual destructor; §5.5; [CG: C.127].

[19] Use **override** to make overriding explicit in large class hierarchies; §5.3; [CG: C.128].

[20] When designing a class hierarchy, distinguish between implementation inheritance and interface inheritance; §5.5.1; [CG: C.129].

[21] Use dynamic_cast where class hierarchy navigation is unavoidable; §5.5.2; [CG: C.146].
[22] Use dynamic_cast to a reference type when failure to find the required class is considered a failure; §5.5.2; [CG: C.147].
[23] Use dynamic_cast to a pointer type when failure to find the required class is considered a valid alternative; §5.5.2; [CG: C.148].
[24] Use unique_ptr or shared_ptr to avoid forgetting to delete objects created using new; §5.5.3; [CG: C.149].

6

Essential Operations

（基本操作）

When someone says
I want a programming language in which
I need only say what I wish done,
give him a lollipop.
– Alan Perlis

- Introduction
  Essential Operations; Conversions; Member Initializers
- Copy and Move
  Copying Containers; Moving Containers
- Resource Management
- Operator Overloading
- Conventional Operations
  Comparisons; Container Operations; Iterators and "Smart Pointers"; Input and Output Operators; `swap()`; `hash<>`
- User-Defined Literals
- Advice

6.1 Introduction（引言）

Some operations, such as initialization, assignment, copy, and move, are fundamental in the sense that language rules make assumptions about them. Other operations, such as `==` and `<<`, have conventional meanings that are perilous to ignore.

6.1.1 Essential Operations（基本操作）

Constructors, destructors, and copy and move operations for a type are not logically separate. We must define them as a matched set or suffer logical or performance problems. If a class `X` has a destructor that performs a nontrivial task, such as free-store deallocation or lock release, the class is likely to need the full complement of functions:

```
class X {
public:
      X(Sometype);                    // "ordinary constructor": create an object
      X();                            // default constructor
      X(const X&);                    // copy constructor
      X(X&&);                         // move constructor
      X& operator=(const X&); // copy assignment: clean up target and copy
      X& operator=(X&&);              // move assignment: clean up target and move
      ~X();                           // destructor: clean up
      // ...
};
```

There are five situations in which an object can be copied or moved:

- As the source of an assignment
- As an object initializer
- As a function argument
- As a function return value
- As an exception

An assignment uses a copy or move assignment operator. In principle, the other cases use a copy or move constructor. However, a copy or move constructor invocation is often optimized away by constructing the object used to initialize right in the target object. For example:

```
X make(Sometype);
X x = make(value);
```

Here, a compiler will typically construct the **X** from **make()** directly in **x**; thus eliminating ("eliding") a copy.

In addition to the initialization of named objects and of objects on the free store, constructors are used to initialize temporary objects and to implement explicit type conversion.

Except for the "ordinary constructor," these special member functions will be generated by the compiler as needed. If you want to be explicit about generating default implementations, you can:

```
class Y {
public:
      Y(Sometype);
      Y(const Y&) = default;     // I really do want the default copy constructor
      Y(Y&&) = default;          // and the default move constructor
      // ...
};
```

If you are explicit about some defaults, other default definitions will not be generated.

When a class has a pointer member, it is usually a good idea to be explicit about copy and move operations. The reason is that a pointer may point to something that the class needs to **delete**, in which case the default memberwise copy would be wrong. Alternatively, it might point to something that the class must *not* **delete**. In either case, a reader of the code would like to know. For an example, see §6.2.1.

A good rule of thumb (sometimes called *the rule of zero*) is to either define all of the essential operations or none (using the default for all). For example:

```
struct Z {
      Vector v;
      string s;
};

Z z1;            // default initialize z1.v and z1.s
Z z2 = z1;       // default copy z1.v and z1.s
```

Here, the compiler will synthesize memberwise default construction, copy, move, and destructor as needed, and all with the correct semantics.
To complement =default, we have =delete to indicate that an operation is not to be generated. A base class in a class hierarchy is the classic example where we don't want to allow a memberwise copy. For example:

```
class Shape {
public:
      Shape(const Shape&) =delete;                    // no copying
      Shape& operator=(const Shape&) =delete;
      // ...
};

void copy(Shape& s1, const Shape& s2)
{
      s1 = s2;   // error: Shape copy is deleted
}
```

A =delete makes an attempted use of the deleted function a compile-time error; =delete can be used to suppress any function, not just essential member functions.

6.1.2 Conversions（转换）

A constructor taking a single argument defines a conversion from its argument type. For example, complex (§5.2.1) provides a constructor from a double:

```
complex z1 = 3.14;  // z1 becomes {3.14,0.0}
complex z2 = z1*2;  // z2 becomes z1*{2.0,0} == {6.28,0.0}
```

This implicit conversion is sometimes ideal, but not always. For example, Vector (§5.2.2) provides a constructor from an int:

```
Vector v1 = 7;  // OK: v1 has 7 elements
```

This is typically considered unfortunate, and the standard-library vector does not allow this int-to-vector "conversion."

The way to avoid this problem is to say that only explicit "conversion" is allowed; that is, we can define the constructor like this:

```
class Vector {
public:
    explicit Vector(int s);     // no implicit conversion from int to Vector
    // ...
};
```

That gives us:

```
Vector v1(7);   // OK: v1 has 7 elements
Vector v2 = 7;  // error: no implicit conversion from int to Vector
```

When it comes to conversions, more types are like **Vector** than are like **complex**, so use **explicit** for constructors that take a single argument unless there is a good reason not to.

6.1.3 Member Initializers（成员初始值设定项）

When a data member of a class is defined, we can supply a default initializer called a *default member initializer*. Consider a revision of **complex** (§5.2.1):

```
class complex {
    double re = 0;
    double im = 0; // representation: two doubles with default value 0.0
public:
    complex(double r, double i) :re{r}, im{i} {}    // construct complex from two scalars: {r,i}
    complex(double r) :re{r} {}                     // construct complex from one scalar: {r,0}
    complex() {}                                    // default complex: {0,0}
    // ...
}
```

The default value is used whenever a constructor doesn't provide a value. This simplifies code and helps us to avoid accidentally leaving a member uninitialized.

6.2 Copy and Move（拷贝和移动）

By default, objects can be copied. This is true for objects of user-defined types as well as for built-in types. The default meaning of copy is memberwise copy: copy each member. For example, using **complex** from §5.2.1:

```
void test(complex z1)
{
    complex z2 {z1};    // copy initialization
    complex z3;
    z3 = z2;            // copy assignment
    // ...
}
```

Now **z1**, **z2**, and **z3** have the same value because both the assignment and the initialization copied both members.

When we design a class, we must always consider if and how an object might be copied. For simple concrete types, memberwise copy is often exactly the right semantics for copy. For some sophisticated concrete types, such as **Vector**, memberwise copy is not the right semantics for copy; for abstract types it almost never is.

6.2.1 Copying Containers （拷贝容器）

When a class is a *resource handle* – that is, when the class is responsible for an object accessed through a pointer – the default memberwise copy is typically a disaster. Memberwise copy would violate the resource handle's invariant (§4.3). For example, the default copy would leave a copy of a **Vector** referring to the same elements as the original:

```
void bad_copy(Vector v1)
{
        Vector v2 = v1;         // copy v1's representation into v2
        v1[0] = 2;              // v2[0] is now also 2!
        v2[1] = 3;              // v1[1] is now also 3!
}
```

Assuming that **v1** has four elements, the result can be represented graphically like this:

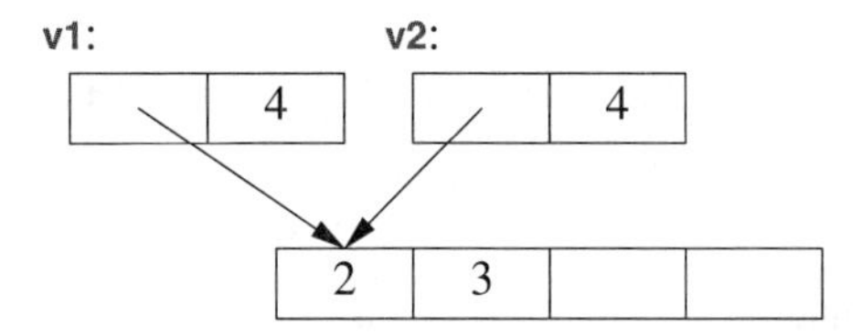

Fortunately, the fact that **Vector** has a destructor is a strong hint that the default (memberwise) copy semantics is wrong and the compiler should at least warn against this example. We need to define better copy semantics.

Copying of an object of a class is defined by two members: a *copy constructor* and a *copy assignment*:

```
class Vector {
public:
        Vector(int s);                          // constructor: establish invariant, acquire resources
        ~Vector() { delete[] elem; }            // destructor: release resources

        Vector(const Vector& a);                // copy constructor
        Vector& operator=(const Vector& a);     // copy assignment

        double& operator[](int i);
        const double& operator[](int i) const;

        int size() const;
private:
        double* elem; // elem points to an array of sz doubles
        int sz;
};
```

A suitable definition of a copy constructor for **Vector** allocates the space for the required number of elements and then copies the elements into it so that after a copy each **Vector** has its own copy of the elements:

```
Vector::Vector(const Vector& a)    // copy constructor
      :elem{new double[a.sz]},     // allocate space for elements
       sz{a.sz}
{
      for (int i=0; i!=sz; ++i)    // copy elements
            elem[i] = a.elem[i];
}
```

The result of the **v2=v1** example can now be presented as:

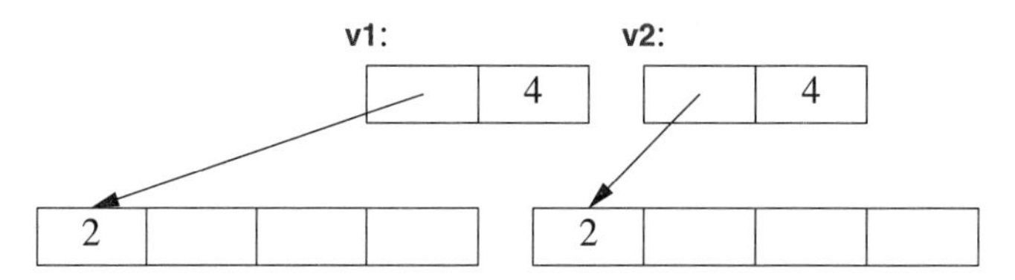

Of course, we need a copy assignment in addition to the copy constructor:

```
Vector& Vector::operator=(const Vector& a)         // copy assignment
{
      double* p = new double[a.sz];
      for (int i=0; i!=a.sz; ++i)
            p[i] = a.elem[i];
      delete[] elem;          // delete old elements
      elem = p;
      sz = a.sz;
      return *this;
}
```

The name **this** is predefined in a member function and points to the object for which the member function is called.

The elements were copied before the old elements were deleted so that if something goes wrong with the element copy and an exception is thrown the old value of the **Vector** is preserved.

6.2.2 Moving Containers（移动容器）

We can control copying by defining a copy constructor and a copy assignment, but copying can be costly for large containers. We avoid the cost of copying when we pass objects to a function by using references, but we can't return a reference to a local object as the result (the local object would be destroyed by the time the caller got a chance to look at it). Consider:

```
Vector operator+(const Vector& a, const Vector& b)
{
      if (a.size()!=b.size())
            throw Vector_size_mismatch{};
```

```
        Vector res(a.size());

        for (int i=0; i!=a.size(); ++i)
              res[i]=a[i]+b[i];
        return res;
}
```

Returning from a + involves copying the result out of the local variable res and into some place where the caller can access it. We might use this + like this:

```
void f(const Vector& x, const Vector& y, const Vector& z)
{
        Vector r;
        // ...
        r = x+y+z;
        // ...
}
```

That would be copying a Vector at least twice (one for each use of the + operator). If a Vector is large, say, 10,000 doubles, that could be embarrassing. The most embarrassing part is that res in operator+() is never used again after the copy. We didn't really want a copy; we just wanted to get the result out of a function: we wanted to *move* a Vector rather than *copy* it. Fortunately, we can state that intent:

```
class Vector {
        // ...

        Vector(const Vector& a);                    // copy constructor
        Vector& operator=(const Vector& a);         // copy assignment

        Vector(Vector&& a);                         // move constructor
        Vector& operator=(Vector&& a);              // move assignment
};
```

Given that definition, the compiler will choose the *move constructor* to implement the transfer of the return value out of the function. This means that r=x+y+z will involve no copying of Vectors. Instead, Vectors are just moved.

As is typical, Vector's move constructor is trivial to define:

```
Vector::Vector(Vector&& a)
        :elem{a.elem},              // "grab the elements" from a
         sz{a.sz}
{
        a.elem = nullptr;           // now a has no elements
        a.sz = 0;
}
```

The && means "rvalue reference" and is a reference to which we can bind an rvalue. The word "rvalue" is intended to complement "lvalue," which roughly means "something that can appear on the left-hand side of an assignment" [Stroustrup,2010]. So an rvalue is – to a first approximation – a value that you can't assign to, such as an integer returned by a function call. Thus, an rvalue reference is a reference to something that *nobody else* can assign to, so we can safely "steal" its value. The **res** local variable in **operator+()** for **Vectors** is an example.

A move constructor does *not* take a **const** argument: after all, a move constructor is supposed to remove the value from its argument. A *move assignment* is defined similarly.

A move operation is applied when an rvalue reference is used as an initializer or as the right-hand side of an assignment.

After a move, the moved-from object should be in a state that allows a destructor to be run. Typically, we also allow assignment to a moved-from object. The standard-library algorithms (Chapter 13) assume that. Our **Vector** does that.

Where the programmer knows that a value will not be used again, but the compiler can't be expected to be smart enough to figure that out, the programmer can be specific:

```
Vector f()
{
        Vector x(1000);
        Vector y(2000);
        Vector z(3000);
        z = x;                  // we get a copy (x might be used later in f())
        y = std::move(x);       // we get a move (move assignment)
        // ... better not use x here ...
        return z;               // we get a move
}
```

The standard-library function **move()** doesn't actually move anything. Instead, it returns a reference to its argument from which we may move – an *rvalue reference*; it is a kind of cast (§5.2.3).

Just before the **return** we have:

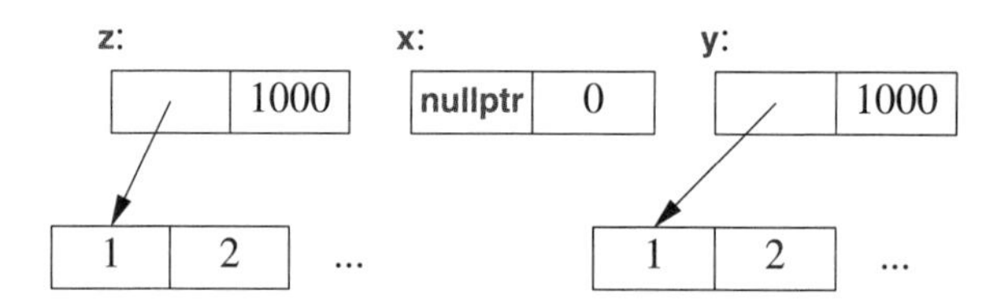

When we return from **f()**, **z** is destroyed after its elements have been moved out of **f()** by the **return**. However, **y**'s destructor will **delete[]** its elements.

The compiler is obliged (by the C++ standard) to eliminate most copies associated with initialization, so move constructors are not invoked as often as you might imagine. This *copy elision* eliminates even the very minor overhead of a move. On the other hand, it is typically not possible to implicitly eliminate copy or move operations from assignments, so move assignments can be critical for performance.

6.3 Resource Management（资源管理）

By defining constructors, copy operations, move operations, and a destructor, a programmer can provide complete control of the lifetime of a contained resource (such as the elements of a container). Furthermore, a move constructor allows an object to move simply and cheaply from one scope to another. That way, objects that we cannot or would not want to copy out of a scope can be simply and cheaply moved out instead. Consider a standard-library **thread** representing a concurrent activity (§18.2) and a **Vector** of a million **double**s. We can't copy the former and don't want to copy the latter.

```
std::vector<thread> my_threads;

Vector init(int n)
{
     thread t {heartbeat};                           // run heartbeat concurrently (in a separate thread)
     my_threads.push_back(std::move(t));             // move t into my_threads (§16.6)
     // ... more initialization ...

     Vector vec(n);
     for (auto& x : vec)
           x = 777;
     return vec;                                     // move vec out of init()
}

auto v = init(1'000'000);          // start heartbeat and initialize v
```

Resource handles, such as **Vector** and **thread**, are superior alternatives to direct use of built-in pointers in many cases. In fact, the standard-library "smart pointers," such as **unique_ptr**, are themselves resource handles (§15.2.1).

I used the standard-library **vector** to hold the **thread**s because we don't get to parameterize our simple **Vector** with an element type until §7.2.

In very much the same way that **new** and **delete** disappear from application code, we can make pointers disappear into resource handles. In both cases, the result is simpler and more maintainable code, without added overhead. In particular, we can achieve *strong resource safety*; that is, we can eliminate resource leaks for a general notion of a resource. Examples are **vector**s holding memory, **thread**s holding system threads, and **fstream**s holding file handles.

In many languages, resource management is primarily delegated to a garbage collector. In C++, you can plug in a garbage collector. However, I consider garbage collection the last choice after cleaner, more general, and better localized alternatives to resource management have been exhausted. My ideal is not to create any garbage, thus eliminating the need for a garbage collector: Do not litter!

Garbage collection is fundamentally a global memory management scheme. Clever implementations can compensate, but as systems are getting more distributed (think caches, multicores, and clusters), locality is more important than ever.

Also, memory is not the only resource. A resource is anything that has to be acquired and (explicitly or implicitly) released after use. Examples are memory, locks, sockets, file handles, and thread handles. Unsurprisingly, a resource that is not just memory is called a *non-memory resource*. A good resource management system handles all kinds of resources. Leaks must be avoided in any long-running system, but excessive resource retention can be almost as bad as a leak. For example, if a system holds on to memory, locks, files, etc. for twice as long, the system needs to be provisioned with potentially twice as many resources.

Before resorting to garbage collection, systematically use resource handles: let each resource have an owner in some scope and by default be released at the end of its owner's scope. In C++, this is known as *RAII* (*Resource Acquisition Is Initialization*) and is integrated with error handling in the form of exceptions. Resources can be moved from scope to scope using move semantics or "smart pointers," and shared ownership can be represented by "shared pointers" (§15.2.1).

In the C++ standard library, RAII is pervasive: for example, memory (string, vector, map, unordered_map, etc.), files (ifstream, ofstream, etc.), threads (thread), locks (lock_guard, unique_lock, etc.), and general objects (through unique_ptr and shared_ptr). The result is implicit resource management that is invisible in common use and leads to low resource retention durations.

6.4 Operator Overloading（操作符重载）

We can give meaning to C++'s operators for user-defined types (§2.4, §5.2.1). That's called *operator overloading* because when used, the correct implementation of an operator must be selected from a set of operators with that same name. For example, our complex + in z1+z2 (§5.2.1) has to be distinguished from the integer + and the floating-point + (§1.4.1).

It is not possible to define new operators, e.g., we cannot define operators ~~, ===, **, $, or unary %. Allowing that would cause as much confusion as good.

It is strongly recommended to define operators with conventional semantics. For example, an operator + that subtracts would do nobody any good.

We can define operators for user-defined types (classes and enumerations):

- Binary arithmetic operators: +, -, *, /, and %
- Binary logical operators: & (bitwise and), | (bitwise or), and ^ (bitwise exclusive or)
- Binary relational operators: ==, !=, <, <=, >, >=, and <=>
- Logical operators: && and ||
- Unary arithmetic and logical operators: +, -, ~, (bitwise complement) and ! (logical negation)
- Assignments: =, +=, *=, etc.
- Increments and decrements: ++ and --
- Pointer operations: ->, unary *, and unary &
- Application (call): ()
- Subscripting: []
- Comma: ,
- Shift: >> and <<

Unfortunately, we cannot define operator dot (.) to get smart references.

An operator can be defined as a member function:

```
class Matrix {
    // ...
    Matrix& operator=(const Matrix& a);        // assign m to *this; return a reference to *this
};
```

This is conventionally done for operators that modify their first operand and is for historical reasons required for =, ->, (), and [].

Alternatively, most operators can be defined as free-standing functions:

```
Matrix operator+(const Matrix& m1, const Matrix& m2);        // assign m1 to m2 and return the sum
```

It is conventional to define operators with symmetric operands as free-standing functions so that both operands are treated identically. To gain good performance from returning a potentially large object, such as a `Matrix`, we rely on move semantics (§6.2.2).

6.5 Conventional Operations（常规操作）

Some operations have conventional meanings when defined for a type. These conventional meanings are often assumed by programmers and libraries (notably, the standard library), so it is wise to conform to them when designing new types for which the operations make sense.

- Comparisons: `==`, `!=`, `<`, `<=`, `>`, `>=`, and `<=>` (§6.5.1)
- Container operations: `size()`, `begin()`, and `end()` (§6.5.2)
- Iterators and “smart pointers”: `->`, `*`, `[]`, `++`, `--`, `+`, `-`, `+=`, and `-=` (§13.3, §15.2.1)
- Function objects: `()` (§7.3.2)
- Input and output operations: `>>` and `<<` (§6.5.4)
- `swap()` (§6.5.5)
- Hash functions: `hash<>` (§6.5.6)

6.5.1 Comparisons (Relational Operators)（比较（关系操作符））

The meaning of the equality comparisons (`==` and `!=`) is closely related to copying. After a copy, the copies should compare equal:

```
X a = something;
X b = a;
assert(a==b);    // if a!=b here, something is very odd (§4.5)
```

When defining `==`, also define `!=` and make sure that `a!=b` means `!(a==b)`.

Similarly, if you define `<`, also define `<=`, `>`, `>=` to make sure that the usual equivalences hold:

- `a<=b` means `(a<b)||(a==b)` and `!(b<a)`.
- `a>b` means `b<a`.
- `a>=b` means `(a>b)||(a==b)` and `!(a<b)`.

To give identical treatment to both operands of a binary operator, such as `==`, it is best defined as a free-standing function in the namespace of its class. For example:

```
namespace NX {
     class X {
          // ...
     };
     bool operator==(const X&, const X&);
     // ...
};
```

The “spaceship operator,” `<=>` is a law onto itself; its rules differ from those for all other operators. In particular, by defining the default `<=>` the other relational operators are implicitly defined:

```
class R {
    // ...
    auto operator<=>(const R& a) const = default;
};

void user(R r1, R r2)
{
    bool b1 = (r1<=>r2) == 0; // r1==r2
    bool b2 = (r1<=>r2) < 0;  // r1<r2
    bool b3 = (r1<=>r2) > 0;  // r1>r2

    bool b4 = (r1==r2);
    bool b5 = (r1<r2);
}
```

Like C's `strcmp()`, `<=>` implements a three-way-comparison. A negative return value means less-than, 0 means equal, and a positive value means greater-than.

If `<=>` is defined as non-default, `==` is not implicitly defined, but `<` and the other relational operators are! For example:

```
struct R2 {
    int m;
    auto operator<=>(const R2& a) const { return a.m == m ? 0 : a.m < m ? -1 : 1; }
};
```

Here, I used the expression form of the **if**-statement : **p?x:y** is an expression that evaluates the condition **p** and if it is true, the value of the **?:** expression is **x** otherwise **y**.

```
void user(R2 r1, R2 r2)
{
    bool b4 = (r1==r2);  // error: no non-default ==
    bool b5 = (r1<r2);   // OK
}
```

This leads to this pattern of definition for nontrivial types:

```
struct R3 { /* ... */ };

auto operator<=>(const R3& a,const R3& b) { /* ... */ }

bool operator==(const R3& a, const R3& b) { /* ... */ }
```

Most standard-library types, such as **string** and **vector**, follow that pattern. The reason is that if a type has more than one element taking part in a comparison, the default `<=>` examines them one at a time yielding a lexicographical order. In such case, it is often worthwhile to provide a separate optimized `==` in addition because `<=>` has to examine all elements to determine all three alternatives. Consider comparing character strings:

```
string s1 = "asdfghjkl";
string s2 = "asdfghjk";

bool b1 = s1==s2;              // false
bool b2 = (s1<=>s2)==0;   // false
```

Using a conventional == we find that the strings are not equal by looking at the number of characters. Using <=>, we have to read all the characters of s2 to find that it is less than s1 and therefore not equal.

There are many more details to operator <=>, but those are primarily of interest to advanced implementors of library facilities concerned with comparisons and sorting beyond the scope of this book. Older code does not use <=>.

6.5.2 Container Operations （容器操作）

Unless there is a really good reason not to, design containers in the style of the standard-library containers (Chapter 12). In particular, make the container resource safe by implementing it as a handle with appropriate essential operations (§6.1.1, §6.2).

The standard-library containers all know their number of elements and we can obtain it by calling size(). For example:

```
for (size_t i = 0; i!=c.size(); ++i)    // size_t is the name of the type returned by a standard-library size()
        c[i] = 0;
```

However, rather than traversing containers using indices from 0 to size(), the standard algorithms (Chapter 13) rely on the notion of *sequences* delimited by pairs of *iterators*:

```
for (auto p = c.begin(); p!=c.end(); ++p)
        *p = 0;
```

Here, c.begin() is an iterator pointing to the first element of c and c.end() points one-beyond-the-last element of c. Like pointers, iterators support ++ to move to the next element and * to access the value of the pointed-to element.

These begin() and end() functions are also used by the implementation of the range-for, so we can simplify loops over a range:

```
for (auto& x : c)
        x= 0;
```

Iterators are used to pass sequences to standard-library algorithms. For example:

```
sort(v.begin(),v.end());
```

This *iterator model* (§13.3) allows for great generality and efficiency. For details and more container operations, see Chapter 12 and Chapter 13.

The begin() and end() can also be defined as free-standing functions; see §7.2. The versions of begin() and end() for const containers are called cbegin() and cend().

6.5.3 Iterators and "smart pointers" （迭代器及智能指针）

User-defined iterators (§13.3) and "smart pointers" (§15.2.1) implement the operators and aspects of a pointer desired for their purpose and often add semantics as needed.

- Access: *, -> (for a class), and [] (for a container)
- Iteration/navigation: ++ (forward), -- (backward) , +=, -=, +, and -
- Copy and/or move: =

6.5.4 Input and Output Operations （输入与输出操作）

For pairs of integers, << means left-shift and >> means right-shift. However, for iostreams, they are the output and input operators, respectively (§1.8, Chapter 11). For details and more I/O operations, see Chapter 11.

6.5.5 swap()

Many algorithms, most notably sort(), use a swap() function that exchanges the values of two objects. Such algorithms generally assume that swap() is very fast and doesn't throw an exception. The standard-library provides a std::swap(a,b) implemented as three move operations (§16.6). If you design a type that is expensive to copy and could plausibly be swapped (e.g., by a sort function), then give it move operations or a swap() or both. Note that the standard-library containers (Chapter 12) and string (§10.2.1) have fast move operations.

6.5.6 hash<>

The standard-library unordered_map<K,V> is a hash table with K as the key type and V as the value type (§12.6). To use a type X as a key, we must define hash<X>. For common types, such as std::string, the standard library defines hash<> for us.

6.6 User-Defined Literals （用户自定义字面量）

One purpose of classes was to enable the programmer to design and implement types to closely mimic built-in types. Constructors provide initialization that equals or exceeds the flexibility and efficiency of built-in type initialization, but for built-in types, we have literals:

- 123 is an int.
- 0xFF00u is an unsigned int.
- 123.456 is a double.
- "Surprise!" is a const char[10].

It can be useful to provide such literals for a user-defined type also. This is done by defining the meaning of a suitable suffix to a literal, so we can get

- "Surprise!"s is a std::string.
- 123s is seconds.
- 12.7i is imaginary so that 12.7i+47 is a complex number (i.e., {47,12.7}).

In particular, we can get these examples from the standard library by using suitable headers and namespaces:

Standard-Library Suffixes for Literals		
<chrono>	std::literals::chrono_literals	h, min, s, ms, us, ns
<string>	std::literals::string_literals	s
<string_view>	std::literals::string_literals	sv
<complex>	std::literals::complex_literals	i, il, if

Literals with user-defined suffixes are called *user-defined literals* or *UDLs*. Such literals are defined using *literal operators*. A literal operator converts a literal of its argument type, followed by a subscript, into its return type. For example, the **i** for **imaginary** suffix might be implemented like this:

```
constexpr complex<double> operator""i(long double arg)     // imaginary literal
{
      return {0,arg};
}
```

Here

- The **operator""** indicates that we are defining a literal operator.
- The **i** after the *literal indicator*, **""**, is the suffix to which the operator gives a meaning.
- The argument type, **long double**, indicates that the suffix (**i**) is being defined for a floating-point literal.
- The return type, **complex<double>**, specifies the type of the resulting literal.

Given that, we can write

```
complex<double> z = 2.7182818+6.283185i;
```

The implementation of the **i** suffix and the **+** are both **constexpr**, so the computation of **z**'s value is done at compile time.

6.7 Advice（建议）

[1] Control construction, copy, move, and destruction of objects; §6.1.1; [CG: R.1].
[2] Design constructors, assignments, and the destructor as a matched set of operations; §6.1.1; [CG: C.22].
[3] Define all essential operations or none; §6.1.1; [CG: C.21].
[4] If a default constructor, assignment, or destructor is appropriate, let the compiler generate it; §6.1.1; [CG: C.20].
[5] If a class has a pointer member, consider if it needs a user-defined or deleted destructor, copy and move; §6.1.1; [CG: C.32] [CG: C.33].
[6] If a class has a user-defined destructor, it probably needs user-defined or deleted copy and move; §6.2.1.
[7] By default, declare single-argument constructors **explicit**; §6.1.2; [CG: C.46].
[8] If a class member has a reasonable default value, provide it as a data member initializer; §6.1.3; [CG: C.48].
[9] Redefine or prohibit copying if the default is not appropriate for a type; §6.1.1; [CG: C.61].

[10] Return containers by value (relying on copy elision and move for efficiency); §6.2.2; [CG: F.20].
[11] Avoid explicit use of **std::copy()**; §16.6; [CG: ES.56].
[12] For large operands, use **const** reference argument types; §6.2.2; [CG: F.16].
[13] Provide strong resource safety; that is, never leak anything that you think of as a resource; §6.3; [CG: R.1].
[14] If a class is a resource handle, it needs a user-defined constructor, a destructor, and non-default copy operations; §6.3; [CG: R.1].
[15] Manage all resources – memory and non-memory – resources using RAII; §6.3; [CG: R.1].
[16] Overload operations to mimic conventional usage; §6.5; [CG: C.160].
[17] If you overload an operator, define all operations that conventionally work together; §6.1.1, §6.5.
[18] If you define <=> for a type as non-default, also define ==; §6.5.1.
[19] Follow the standard-library container design; §6.5.2; [CG: C.100].

7

Templates
（模板）

Your quote here.
– B. Stroustrup

7.1 Introduction（引言）

Someone who wants a vector is unlikely always to want a vector of **double**s. A vector is a general concept, independent of the notion of a floating-point number. Consequently, the element type of a vector ought to be represented independently. A *template* is a class or a function that we parameterize with a set of types or values. We use templates to represent ideas that are best understood as something general from which we can generate specific types and functions by specifying arguments, such as **double** as **vector**'s element type.

This chapter focuses on language mechanisms. Chapter 8 follows up with programming techniques, and the library chapters (Chapters 10–18) offer many examples.

7.2 Parameterized Types（参数化类型）

We can generalize our vector-of-doubles type (§5.2.2) to a vector-of-anything type by making it a **template** and replacing the specific type **double** with a type parameter. For example:

```
template<typename T>
class Vector {
private:
      T* elem;  // elem points to an array of sz elements of type T
      int sz;
public:
      explicit Vector(int s);            // constructor: establish invariant, acquire resources
      ~Vector() { delete[] elem; }       // destructor: release resources

      // ... copy and move operations ...

      T& operator[](int i);                    // for non-const Vectors
      const T& operator[](int i) const;        // for const Vectors (§5.2.1)
      int size() const { return sz; }
};
```

The **template<typename T>** prefix makes **T** a type parameter of the declaration it prefixes. It is C++'s version of the mathematical "for all T" or more precisely "for all types T." If you want the mathematical "for all T, such that P(T)," you use concepts (§7.2.1, §8.2). Using **class** to introduce a type parameter is equivalent to using **typename**, and in older code we often see **template<class T>** as the prefix.

The member functions can be defined similarly:

```
template<typename T>
Vector<T>::Vector(int s)
{
      if (s<0)
            throw length_error{"Vector constructor: negative size"};
      elem = new T[s];
      sz = s;
}

template<typename T>
const T& Vector<T>::operator[](int i) const
{
      if (i<0 || size()<=i)
            throw out_of_range{"Vector::operator[]"};
      return elem[i];
}
```

Given these definitions, we can define **Vectors** like this:

```
Vector<char> vc(200);          // vector of 200 characters
Vector<string> vs(17);         // vector of 17 strings
Vector<list<int>> vli(45);     // vector of 45 lists of integers
```

The >> in Vector<list<int>> terminates the nested template arguments; it is not a misplaced input operator.

We can use Vectors like this:

```
void write(const Vector<string>& vs)          // Vector of some strings
{
    for (int i = 0; i!=vs.size(); ++i)
        cout << vs[i] << '\n';
}
```

To support the range-for loop for our Vector, we must define suitable begin() and end() functions:

```
template<typename T>
T* begin(Vector<T>& x)
{
    return &x[0];    // pointer to first element or to one-past-the-last element
}

template<typename T>
T* end(Vector<T>& x)
{
    return &x[0]+x.size();          // pointer to one-past-the-last element
}
```

Given those, we can write:

```
void write2(Vector<string>& vs)     // Vector of some strings
{
    for (auto& s : vs)
        cout << s << '\n';
}
```

Similarly, we can define lists, vectors, maps (that is, associative arrays), unordered maps (that is, hash tables), etc., as templates (Chapter 12).

Templates are a compile-time mechanism, so their use incurs no run-time overhead compared to hand-crafted code. In fact, the code generated for Vector<double> is identical to the code generated for the version of Vector from Chapter 5. Furthermore, the code generated for the standard-library vector<double> is likely to be better (because more effort has gone into its implementation).

A template plus a set of template arguments is called an *instantiation* or a *specialization*. Late in the compilation process, at *instantiation time*, code is generated for each instantiation used in a program (§8.5).

7.2.1 Constrained Template Arguments（受限模板参数）

Most often, a template will make sense only for template arguments that meet certain criteria. For example, a Vector typically offers a copy operation, and if it does, it must require that its elements are copyable. That is, we must require that Vector's template argument is not just a typename but an Element where "Element" specifies the requirements of a type that can be an element:

```
template<Element T>
class Vector {
private:
    T* elem;  // elem points to an array of sz elements of type T
    int sz;
    // ...
};
```

This template<Element T> prefix is C++'s version of mathematic's "for all T such that Element(T)"; that is, **Element** is a predicate that checks whether **T** has all the properties that a **Vector** requires. Such a predicate is called a *concept* (§8.2). A template argument for which a concept is specified is called a *constrained argument* and a template for which an argument is constrained is called a *constrained template*.

The requirements for the type of a standard-library element are a bit complicated (§12.2), but for our simple **Vector**, **Element** could be something like the standard-library concept **copyable** (§14.5).

It is a compile-time error to try to use a template with a type that does not meet its requirements. For example:

```
Vector<int> v1;      // OK: we can copy an int
Vector<thread> v2;  // error: we can't copy a standard thread (§18.2)
```

Thus, concepts lets the compiler to do type checking at the point of use, giving better error messages far earlier than is possible with unconstrained template arguments. C++ did not officially support concepts before C++20, so older code uses unconstrained template arguments and leaves requirements to documentation. However, the code generated from templates is type checked so that even unconstrained template code is as type safe as handwritten code. For unconstrained parameters, that type check cannot be done until the types of all entities involved are available, so it can occur unpleasantly late in the compilation process, at instantiation time (§8.5), and the error messages are often atrocious.

Concept checking is a purely compile-time mechanism and the code generated is as good as that from unconstrained templates.

7.2.2 Value Template Arguments （模板值参数）

In addition to type arguments, a template can take value arguments. For example:

```
template<typename T, int N>
struct Buffer {
    constexpr int size() { return N; }
    T elem[N];
    // ...
};
```

Value arguments are useful in many contexts. For example, **Buffer** allows us to create arbitrarily sized buffers with no use of the free store (dynamic memory):

```
Buffer<char,1024> glob;  // global buffer of characters (statically allocated)

void fct()
{
     Buffer<int,10> buf;  // local buffer of integers (on the stack)
     // ...
}
```

Unfortunately, for obscure technical reasons, a string literal cannot yet be a template value argument. However, in some contexts, the ability to parameterize with string values is critically important. Fortunately, we can use an array holding the characters of a string:

```
template<char* s>
void outs() { cout << s; }

char arr[] = "Weird workaround!";

void use()
{
     outs<"straightforward use">();          // error (for now)
     outs<arr>();                            // writes: Weird workaround!
}
```

In C++, there is usually a workaround; we don't need direct support for every use case.

7.2.3 Template Argument Deduction （模板参数推导）

When defining a type as an instantiation of a template we must specify its template arguments. Consider using the standard-library template pair:

```
pair<int,double> p = {1, 5.2};
```

Having to specify the template argument types can be tedious. Fortunately, in many contexts, we can simply let pair's constructor deduce the template arguments from an initializer:

```
pair p = {1, 5.2};          // p is a pair<int,double>
```

Containers provide another example:

```
template<typename T>
class Vector {
public:
     Vector(int);
     Vector(initializer_list<T>);       // initializer-list constructor
     // ...
};

Vector v1 {1, 2, 3};          // deduce v1's element type from the initializer element type: int
Vector v2 = v1;               // deduce v2's element type from v1's element type: int
```

```
auto p = new Vector{1, 2, 3};   // p is a Vector<int>*

Vector<int> v3(1);   // here we need to be explicit about the element type (no element type is mentioned)
```

Clearly, this simplifies notation and can eliminate annoyances caused by mistyping redundant template argument types. However, it is not a panacea. Like all other powerful mechanisms, deduction can cause surprises. Consider:

```
Vector<string> vs {"Hello", "World"};     // OK: Vector<string>
Vector vs1 {"Hello", "World"};            // OK: deduces to Vector<const char*> (Surprise?)
Vector vs2 {"Hello"s, "World"s};          // OK: deduces to Vector<string>
Vector vs3 {"Hello"s, "World"};           // error: the initializer list is not homogenous
Vector<string> vs4 {"Hello"s, "World"};   // OK: the element type is explicit
```

The type of a C-style string literal is **const char*** (§1.7.1). If that is not what is intended for **vs1**, we must be explicit about the element type or use the **s** suffix to make it a proper **string** (§10.2).

If elements of an initializer list have differing types, we cannot deduce a unique element type, so we get an ambiguity error.

Sometimes, we need to resolve an ambiguity. For example, the standard-library **vector** has a constructor that takes a pair of iterators delimiting a sequence and also an initializer constructor that can take a pair of values. Consider:

```
template<typename T>
class Vector {
public:
     Vector(initializer_list<T>);     // initializer-list constructor

     template<typename Iter>
          Vector(Iter b, Iter e);     // [b:e) iterator-pair constructor

     struct iterator { using value_type = T; /* ... */ };
     iterator begin();

     // ...
};

Vector v1 {1, 2, 3, 4, 5};                 // element type is int
Vector v2(v1.begin(),v1.begin()+2);        // a pair of iterators or a pair of values (of type iterator)?
Vector v3(9,17);        // error: ambiguous
```

We could resolve this using concepts (§8.2), but the standard library and many other important bodies of code were written decades before we had language support for concepts. For those, we need a way of saying "a pair of values of the same type should be considered iterators." Adding a *deduction guide* after the declaration of **Vector** does exactly that:

```
template<typename Iter>
     Vector(Iter,Iter) -> Vector<typename Iter::value_type>;
```

Now we have:

```
Vector v1 {1, 2, 3, 4, 5};                // element type is int
Vector v2(v1.begin(),v1.begin()+2);       // pair-of-iterators: element type is int
Vector v3 {v1.begin(),v1.begin()+2};      // element type is Vector2::iterator
```

The {} initializer syntax always prefers the **initializer_list** constructor (if present), so **v3** is a vector of iterators: **Vector<Vector<int>::iterator>**.

The () initialization syntax (§12.2) is conventional for when we don't want an **initializer_list**.

The effects of deduction guides are often subtle, so it is best to design class templates so that deduction guides are not needed.

People who like acronyms refer to "class template argument deduction" as *CTAD*.

7.3 Parameterized Operations（参数化操作）

Templates have many more uses than simply parameterizing a container with an element type. In particular, they are extensively used for parameterization of both types and algorithms in the standard library (§12.8, §13.5).

There are three ways of expressing an operation parameterized by types or values:

- A function template
- A function object: an object that can carry data and be called like a function
- A lambda expression: a shorthand notation for a function object

7.3.1 Function Templates（模板函数）

We can write a function that calculates the sum of the element values of any sequence that a range-**for** can traverse (e.g., a container) like this:

```
template<typename Sequence, typename Value>
Value sum(const Sequence& s, Value v)
{
     for (auto x : s)
          v+=x;
     return v;
}
```

The **Value** template argument and the function argument **v** are there to allow the caller to specify the type and initial value of the accumulator (the variable in which to accumulate the sum):

```
void user(Vector<int>& vi, list<double>& ld, vector<complex<double>>& vc)
{
     int x = sum(vi,0);                        // the sum of a vector of ints (add ints)
     double d = sum(vi,0.0);                   // the sum of a vector of ints (add doubles)
     double dd = sum(ld,0.0);                  // the sum of a list of doubles
     auto z = sum(vc,complex{0.0,0.0});        // the sum of a vector of complex<double>s
}
```

The point of adding **int**s in a **double** would be to gracefully handle a sum larger than the largest **int**. Note how the types of the template arguments for **sum<Sequence,Value>** are deduced from the function arguments. Fortunately, we do not need to explicitly specify those types.

This **sum()** is a simplified version of the standard-library **accumulate()** (§17.3).

A function template can be a member function, but not a **virtual** member. The compiler would not know all instantiations of such a template in a program, so it could not generate a **vtbl** (§5.4).

7.3.2 Function Objects（函数对象）

One particularly useful kind of template is the *function object* (sometimes called a *functor*), which is used to define objects that can be called like functions. For example:

```
template<typename T>
class Less_than {
      const T val;    // value to compare against
public:
      Less_than(const T& v) :val{v} { }
      bool operator()(const T& x) const { return x<val; } // call operator
};
```

The function called **operator()** implements the *application operator*, **()**, also called "function call" or just "call."

We can define named variables of type **Less_than** for some argument type:

```
Less_than lti {42};                 // lti(i) will compare i to 42 using < (i<42)
Less_than lts {"Backus"s};          // lts(s) will compare s to "Backus" using < (s<"Backus")
Less_than<string> lts2 {"Naur"};    // "Naur" is a C-style string, so we need <string> to get the right <
```

We can call such an object, just as we call a function:

```
void fct(int n, const string& s)
{
      bool b1 = lti(n);      // true if n<42
      bool b2 = lts(s);      // true if s<"Backus"
      // ...
}
```

Function objects are widely used as arguments to algorithms. For example, we can count the occurrences of values for which a predicate returns **true**:

```
template<typename C, typename P>
int count(const C& c, P pred)      // assume that C is a container and P is a predicate on its elements
{
      int cnt = 0;
      for (const auto& x : c)
            if (pred(x))
                  ++cnt;
      return cnt;
}
```

This is a simplified version of the standard-library **count_if** algorithm (§13.5).

Given concepts (§8.2), we can formalize **count()**'s assumptions about its argument and check them at compile time.

A *predicate* is something that we can invoke to return **true** or **false**. For example:

```
void f(const Vector<int>& vec, const list<string>& lst, int x, const string& s)
{
    cout << "number of values less than " << x << ": " << count(vec,Less_than{x}) << '\n';
    cout << "number of values less than " << s << ": " << count(lst,Less_than{s}) << '\n';
}
```

Here, Less_than{x} constructs an object of type Less_than<int>, for which the call operator compares to the value of the int called x; Less_than{s} constructs an object that compares to the value of the string called s.

The beauty of function objects is that they carry the value to be compared against with them. We don't have to write a separate function for each value (and each type), and we don't have to introduce nasty global variables to hold values. Also, for a simple function object like Less_than, inlining is simple, so a call of Less_than is far more efficient than an indirect function call. The ability to carry data plus their efficiency makes function objects particularly useful as arguments to algorithms.

Function objects used to specify the meaning of key operations of a general algorithm (such as Less_than for count()) are sometimes referred to as *policy objects*.

7.3.3 Lambda Expressions （匿名函数表达式）

In §7.3.2, we defined Less_than separately from its use. That can be inconvenient. Consequently, there is a notation for implicitly generating function objects:

```
void f(const Vector<int>& vec, const list<string>& lst, int x, const string& s)
{
    cout << "number of values less than " << x
         << ": " << count(vec,[&](int a){ return a<x; })
         << '\n';

    cout << "number of values less than " << s
         << ": " << count(lst,[&](const string& a){ return a<s; })
         << '\n';
}
```

The notation [&](int a){ return a<x; } is called a *lambda expression*. It generates a function object similar to Less_than<int>{x}. The [&] is a *capture list* specifying that all local names used in the lambda body (such as x) will be accessed through references. Had we wanted to "capture" only x, we could have said so: [&x]. Had we wanted to give the generated object a copy of x, we could have said so: [x]. Capture nothing is [ ], capture all local names used by reference is [&], and capture all local names used by value is [=].

For a lambda defined within a member function, [this] captures the current object by reference so that we can refer to class members. If we want a copy of the current object, we say [*this].

If we want to capture several specific objects, we can list them. The use of [i,this] in the use of expect() (§4.5) is an example.

7.3.3.1 Lamdas as function arguments（匿名函数作为函数参数）

Using lambdas can be convenient and terse, but also obscure. For nontrivial actions (say, more than a simple expression), I prefer to name the operation so as to more clearly state its purpose and to make it available for use in several places in a program.

In §5.5.3, we noted the annoyance of having to write many functions to perform operations on elements of **vector**s of pointers and **unique_ptr**s, such as **draw_all()** and **rotate_all()**. Function objects (in particular, lambdas) can help by allowing us to separate the traversal of the container from the specification of what is to be done with each element.

First, we need a function that applies an operation to each object pointed to by the elements of a container of pointers:

```
template<typename C, typename Oper>
void for_each(C& c, Oper op)        // assume that C is a container of pointers (see also §8.2.1)
{
      for (auto& x : c)
            op(x);          // pass op() a reference to each element pointed to
}
```

This is a simplified version of the standard-library **for_each** algorithm (§13.5).

Now, we can write a version of **user()** from §5.5 without writing a set of **_all** functions:

```
void user()
{
      vector<unique_ptr<Shape>> v;
      while (cin)
            v.push_back(read_shape(cin));
      for_each(v,[](unique_ptr<Shape>& ps){ ps->draw(); });          // draw_all()
      for_each(v,[](unique_ptr<Shape>& ps){ ps->rotate(45); });     // rotate_all(45)
}
```

I pass the **unique_ptr<Shape>**s to the lambdas by reference. That way **for_each()** doesn't have to deal with lifetime issues.

Like a function, a lambda can be generic. For example:

```
template<class S>
void rotate_and_draw(vector<S>& v, int r)
{
      for_each(v,[](auto& s){ s->rotate(r); s->draw(); });
}
```

Here, like in variable declarations, **auto** means that a value of any type is accepted as an initializer (an argument is considered to initialize the formal parameter in a call). This makes a lambda with an **auto** parameter a template, a *generic lambda*. When needed, we can constrain the parameter with a concept (§8.2). For example, we could define **Pointer_to_class** to require * and -> and write:

```
for_each(v,[](Pointer_to_class auto& s){ s->rotate(r); s->draw(); });
```

We can call this generic **rotate_and_draw()** with any container of objects that you can **draw()** and **rotate()**. For example:

```
void user()
{
	vector<unique_ptr<Shape>> v1;
	vector<Shape*> v2;
	// ...
	rotate_and_draw(v1,45);
	rotate_and_draw(v2,90);
}
```

For even tighter checking, we could define a **Pointer_to_Shape** concept specifying the properties we want for a type to be useable as a shape. That would allow us to use shapes that weren't derived from class **Shape**.

7.3.3.2 Lambdas for initialization（匿名函数与初始化）

Using a lambda, we can turn any statement into an expression. This is mostly used to provide an operation to compute a value as an argument value, but the ability is general. Consider a complicated initialization:

```
enum class Init_mode { zero, seq, cpy, patrn };	// initializer alternatives

void user(Init_mode m, int n, vector<int>& arg, Iterator p, Iterator q)
{
	vector<int> v;

	// messy initialization code:

	switch (m) {
	case zero:
		v = vector<int>(n);	// n elements initialized to 0
		break;
	case cpy:
		v = arg;
		break;
	};

	// ...

	if (m == seq)
		v.assign(p,q);		// copy from sequence [p:q)

	// ...
}
```

This is a stylized example, but unfortunately not atypical. We need to select among a set of alternatives for initializing a data structure (here **v**) and we need to do different computations for different alternatives. Such code is often messy, deemed essential "for efficiency," and a source of bugs:

- The variable could be used before it gets its intended value.

- The "initialization code" could be mixed with other code, making it hard to comprehend.
- When "initialization code" is mixed with other code it is easier to forget a case.
- This isn't initialization, it's assignment (§1.9.2).

Instead, we could convert it to a lambda used as an initializer:

```
void user(Init_mode m, int n, vector<int>& arg, Iterator p, Iterator q)
{
	vector<int> v = [&] {
		switch (m) {
		case zero:	return vector<int>(n);		// n elements initialized to 0
		case seq:	return vector<int>{p,q};	// copy from sequence [p:q]
		case cpy:	return arg;
		}
	}();

	// ...
}
```

I still "forgot" a **case**, but now that's more easily spotted. In many cases, a compiler will spot the problem and warn.

7.3.3.3 Finally（作用域终结函数）

Destructors offer a general and implicit mechanism for cleaning up after use of an object (RAII; §6.3), but what if we need to do some cleanup that is not associated with a single object, or with an object that does not have a destructor (e.g., because it is a type shared with a C program)? We can define a function, **finally()** that takes an action to be executed on the exit from the scope

```
void old_style(int n)
{
	void* p = malloc(n*sizeof(int));	// C-style
	auto act = finally([&]{free(p);});	// call the lambda upon scope exit
	// ...
}	// p is implicitly freed upon scope exit
```

This is ad hoc, but far better than trying to correctly and consistently call **free(p)** on all exits from the function.

The **finally()** function is trivial:

```
template <class F>
[[nodiscard]] auto finally(F f)
{
	return Final_action{f};
}
```

I used the attribute **[[nodiscard]]** to ensure that users do not forget to copy a generated **Final_action** into the scope for which its action is intended.

The class **Final_action** that supplies the necessary destructor can look like this:

```
template <class F>
struct Final_action {
      explicit Final_action(F f) :act(f) {}
      ˜Final_action() { act(); }
      F act;
};
```

There is a **finally()** in the Core Guidelines Support Library (the GSL) and a proposal for a more elaborate **scope_exit** mechanism for the standard library.

7.4 Template Mechanisms （模板机制）

To define good templates, we need some supporting language facilities:

- Values dependent on a type: *variable templates* (§7.4.1).
- Aliases for types and templates: *alias templates* (§7.4.2).
- A compile-time selection mechanism: **if constexpr** (§7.4.3).
- A compile-time mechanism to inquire about properties of types and expressions: **requires**-expressions (§8.2.3).

In addition, **constexpr** functions (§1.6) and **static_asserts** (§4.5.2) often take part in template design and use.

These basic mechanisms are primarily tools for building general, foundational abstractions.

7.4.1 Variable Templates （模板变量）

When we use a type, we often want constants and values of that type. This is of course also the case when we use a class template: when we define a **C<T>**, we often want constants and variables of type **C<T>** and other types depending on **T**. Here is an example from a fluid dynamic simulation [Garcia,2015]:

```
template <class T>
      constexpr T viscosity = 0.4;

template <class T>
      constexpr space_vector<T> external_acceleration = { T{}, T{-9.8}, T{} };

auto vis2 = 2*viscosity<double>;
auto acc = external_acceleration<float>;
```

Here, **space_vector** is a three-dimensional vector.

Curiously enough, most variable templates seem to be constants. But then, so are many variables. Terminology hasn't kept up with our notions of immutability.

Naturally, we can use arbitrary expressions of suitable types as initializers. Consider:

```
template<typename T, typename T2>
constexpr bool Assignable = is_assignable<T&,T2>::value;    // is_assignable is a type trait (§16.4.1)
```

```
template<typename T>
void testing()
{
      static_assert(Assignable<T&,double>, "can't assign a double to a T");
      static_assert(Assignable<T&,string>, "can't assign a string to a T");
}
```

After some significant mutations, this idea becomes the heart of concept definitions (§8.2).

The standard library uses variable templates to provide mathematical constants, such as pi and log2e (§17.9).

7.4.2 Aliases （别名）

Surprisingly often, it is useful to introduce a synonym for a type or a template. For example, the standard header <cstddef> contains a definition of the alias size_t, maybe:

```
using size_t = unsigned int;
```

The actual type named size_t is implementation-dependent, so in another implementation size_t may be an unsigned long. Having the alias size_t allows the programmer to write portable code.

It is very common for a parameterized type to provide an alias for types related to their template arguments. For example:

```
template<typename T>
class Vector {
public:
      using value_type = T;
      // ...
};
```

In fact, every standard-library container provides value_type as the name for the type of its elements (Chapter 12). This allows us to write code that will work for every container that follows this convention. For example:

```
template<typename C>
using Value_type = C::value_type;          // the type of C's elements

template<typename Container>
void algo(Container& c)
{
      Vector<Value_type<Container>> vec;          // keep results here
      // ...
}
```

This Value_type is a simplified version of the standard library range_value_t (§16.4.4). The aliasing mechanism can be used to define a new template by binding some or all template arguments. For example:

```
template<typename Key, typename Value>
class Map {
	// ...
};

template<typename Value>
using String_map = Map<string,Value>;

String_map<int> m;	// m is a Map<string,int>
```

7.4.3 Compile-Time if （编译时 if）

Consider writing an operation that can be implemented using one of two functions **slow_and_safe(T)** or **simple_and_fast(T)**. Such problems abound in foundational code where generality and optimal performance are essential. If a class hierarchy is involved, a base class can provide the **slow_and_safe** general operation and a derived class can override with a **simple_and_fast** implementation.

Alternatively, we can use a compile-time **if**:

```
template<typename T>
void update(T& target)
{
	// ...
	if constexpr(is_trivially_copyable_v<T>)
		simple_and_fast(target);	// for "plain old data"
	else
		slow_and_safe(target);	// for more complex types
	// ...
}
```

The **is_trivially_copyable_v<T>** is a type predicate (§16.4.1) that tells us if a type can be trivially copied.

Only the selected branch of an **if constexpr** is checked by the compiler. This solution offers optimal performance and locality of the optimization.

Importantly, **if constexpr** is not a text-manipulation mechanism and cannot be used to break the usual rules of grammar, type, and scope. For example, here is a naive and failed attempt to conditionally wrap a call in a **try**-block:

```
template<typename T>
void bad(T arg)
{
	if constexpr(!is_trivially_copyable_v<T>)
		try {	// Oops, the if extends beyond this line

	g(arg);
```

```
        if constexpr(!is_trivially_copyable_v<T>)
            } catch(...) { /* ... */ }              // syntax error
}
```

Allowing such text manipulation could seriously compromise readability of code and create problems for tools relying on modern program representation techniques (such as "abstract syntax trees").

Many such attempted hacks are also unnecessary because cleaner solutions that do not violate scope rules are available. For example:

```
template<typename T>
void good(T arg)
{
    if constexpr (is_trivially_copyable_v<T>)
        g(arg);
    else
        try {
            g(arg);
        }
        catch (...) { /* ... */ }
}
```

7.5 Advice（建议）

[1] Use templates to express algorithms that apply to many argument types; §7.1; [CG: T.2].
[2] Use templates to express containers; §7.2; [CG: T.3].
[3] Use templates to raise the level of abstraction of code; §7.2; [CG: T.1].
[4] Templates are type safe, but for unconstrained templates checking happens too late; §7.2.
[5] Let constructors or function templates deduce class template argument types; §7.2.3.
[6] Use function objects as arguments to algorithms; §7.3.2; [CG: T.40].
[7] Use a lambda if you need a simple function object in one place only; §7.3.2.
[8] A virtual function member cannot be a template member function; §7.3.1.
[9] Use finally() to provide RAII for types without destructors that require "cleanup operations"; §7.3.3.3.
[10] Use template aliases to simplify notation and hide implementation details; §7.4.2.
[11] Use if constexpr to provide alternative implementations without run-time overhead; §7.4.3.

8

Concepts and Generic Programming

（概念和泛型编程）

Programming:
you have to start with interesting algorithms.
– Alex Stepanov

- Introduction
- Concepts
  Use of Concepts; Concept-Based Overloading; Valid Code; Definition of Concepts; Concepts and **auto**; Concepts and Types
- Generic Programming
  Use of Concepts; Abstraction Using Templates
- Variadic Templates
  Fold Expressions; Forwarding Arguments
- Template Compilation Model
- Advice

8.1 Introduction（引言）

What are templates for? In other words, what programming techniques are made effective by templates? Templates offer:

- The ability to pass types (as well as values and templates) as arguments without loss of information. This implies great flexibility in what can be expressed and excellent opportunities for inlining, of which current implementations take great advantage.
- Opportunities to weave together information from different contexts at instantiation time. This implies optimization opportunities.
- The ability to pass values as template arguments. This implies opportunities for compile-time computation.

In other words, templates provide a powerful mechanism for compile-time computation and type manipulation that can lead to very compact and efficient code. Remember that types (classes) can contain both code (§7.3.2) and values (§7.2.2).

The first and most common use of templates is to support *generic programming*, that is, programming focused on the design, implementation, and use of general algorithms. Here, "general" means that an algorithm can be designed to accept a wide variety of types as long as they meet the algorithm's requirements on its arguments. Together with concepts, the template is C++'s main support for generic programming. Templates provide (compile-time) parametric polymorphism.

8.2 Concepts（概念）

Consider the **sum()** from §7.3.1:

```
template<typename Seq, typename Value>
Value sum(Seq s, Value v)
{
	for (const auto& x : s)
		v+=x;
	return v;
}
```

This **sum()** requires that

- its first template argument is some kind of sequence of elements, and
- its second template argument is some kind of number.

To be more specific, **sum()** can be invoked for a pair of arguments:

- A *sequence*, **Seq**, that supports **begin()** and **end()** so that the range-**for** will work (§1.7; §14.1).
- An *arithmetic type*, **Value**, that supports **+=** so that elements of the sequence can be added.

We call such requirements *concepts*.

Examples of types that meet this simplified requirement (and more) for being a sequence (also called a *range*) include the standard-library **vector**, **list**, and **map**. Examples of types that meet this simplified requirement (and more) for being an arithmetic type include **int**, **double**, and **Matrix** (for any reasonable definition of **Matrix**). We could say that the **sum()** algorithm is generic in two dimensions: the type of the data structure used to store elements ("the sequence") and the type of elements.

8.2.1 Use of Concepts（概念的运用）

Most template arguments must meet specific requirements for the template to compile properly and for the generated code to work properly. That is, most templates should be constrained templates (§7.2.1). The type-name introducer **typename** is the least constraining, requiring only that the argument be a type. Usually, we can do better than that. Consider that **sum()** again:

```
template<Sequence Seq, Number Num>
Num sum(Seq s, Num v)
{
	for (const auto& x : s)
		v+=x;
	return v;
}
```

That's much clearer. Once we have defined what the concepts `Sequence` and `Number` mean, the compiler can reject bad calls by looking at `sum()`'s interface only, rather than looking at its implementation. This improves error reporting.

However, the specification of `sum()`'s interface is not complete: I "forgot" to say that we should be able to add elements of a `Sequence` to a `Number`. We can do that:

```
template<Sequence Seq, Number Num>
     requires Arithmetic<range_value_t<Seq>,Num>
Num sum(Seq s, Num n);
```

The `range_value_t` (§16.4.4) of a sequence is the type of the elements in that sequence; it comes from the standard library where it names the type of the elements of a `range` (§14.1). `Arithmetic<X,Y>` is a concept specifying that we can do arithmetic with numbers of types `X` and `Y`. This saves us from accidentally trying to calculate the `sum()` of a `vector<string>` or a `vector<int*>` while still accepting `vector<int>` and `vector<complex<double>>`. Typically, when an algorithm requires arguments of differing types, there is a relationship between those types that it is good to make explicit.

In this example, we needed only `+=`, but for simplicity and flexibility, we should not constrain our template argument too tightly. In particular, we might someday want to express `sum()` in terms of `+` and `=` rather than `+=`, and then we'd be happy that we used a general concept (here, `Arithmetic`) rather than a narrow requirement to "have `+=`."

Partial specifications, as in the first `sum()` using concepts, can be very useful. Unless the specification is complete, some errors will not be found until instantiation time. However, even partial specifications express intent and are essential for smooth incremental development where we don't initially recognize all the requirements we need. With mature libraries of concepts, initial specifications will be close to perfect.

Unsurprisingly, `requires Arithmetic<range_value_t<Seq>,Num>` is called a `requirements`-clause. The `template<Sequence Seq>` notation is simply a shorthand for an explicit use of `requires Sequence<Seq>`. If I liked verbosity, I could equivalently have written

```
template<typename Seq, typename Num>
     requires Sequence<Seq> && Number<Num> && Arithmetic<range_value_t<Seq>,Num>
Num sum(Seq s, Num n);
```

On the other hand, we could also use the equivalence between the two notations to write:

```
template<Sequence Seq, Arithmetic<range_value_t<Seq>> Num>
Num sum(Seq s, Num n);
```

In code bases where we cannot yet use `concepts`, we have to make do with naming conventions and comments, such as:

```
template<typename Sequence, typename Number>
     // requires Arithmetic<range_value_t<Sequence>,Number>
Number sum(Sequence s, Number n);
```

Whatever notation we choose, it is important to design a template with semantically meaningful constraints on its arguments (§8.2.4).

8.2.2 Concept-based Overloading （基于概念的重载）

Once we have properly specified templates with their interfaces, we can overload based on their properties, much as we do for functions. Consider a slightly simplified standard-library function **advance()** that advances an iterator (§13.3):

```
template<forward_iterator Iter>
void advance(Iter p, int n)            // move p n elements forward
{
      while (n--)
            ++p;        // a forward iterator has ++, but not + or +=
}

template<random_access_iterator Iter>
void advance(Iter p, int n)            // move p n elements forward
{
      p+=n;             // a random-access iterator has +=
}
```

The compiler will select the template with the strongest requirements met by the arguments. In this case, a **list** only supplies forward iterators, but a **vector** offers random-access iterators, so we get:

```
void user(vector<int>::iterator vip, list<string>::iterator lsp)
{
      advance(vip,10);      // uses the fast advance()
      advance(lsp,10);      // uses the slow advance()
}
```

Like other overloading, this is a compile-time mechanism implying no run-time cost, and where the compiler does not find a best choice, it gives an ambiguity error. The rules for concept-based overloading are far simpler than the rules for general overloading (§1.3). Consider first a single argument for several alternative functions:

- If the argument doesn't match the concept, that alternative cannot be chosen.
- If the argument matches the concept for just one alternative, that alternative is chosen.
- If arguments from two alternatives match a concept and one is stricter than the other (match all the requirements of the other and more), that alternative is chosen.
- If arguments from two alternatives are equally good matches for a concept, we have an ambiguity.

For an alternative to be chosen it must be

- a match for all of its arguments, and
- at least an equally good match for all arguments as other alternatives, and
- a better match for at least one argument.

8.2.3 Valid Code （有效代码）

The question of whether a set of template arguments offers what a template requires of its template parameters ultimately boils down to whether some expressions are valid.

Using a **requires**-expression, we can check if a set of expressions is valid. For example, we might try to write **advance()** without the use of the standard-library concept **random_access_iterator**:

```
template<forward_iterator Iter>
    requires requires(Iter p, int i) { p[i]; p+i; }     // Iter has subscripting and integer addition
void advance(Iter p, int n)           // move p n elements forward
{
    p+=n;
}
```

No, that **requires requires** is not a typo. The first **requires** starts the **requirements**-clause and the second **requires** starts the **requires**–expression

```
    requires(Iter p, int i) { p[i]; p+i; }
```

A **requires**–expression is a predicate that is **true** if the statements in it are valid code and **false** if not.

I consider **requires**-expressions the assembly code of generic programming. Like ordinary assembly code, **requires**-expressions are extremely flexible and impose no programming discipline. In some form or other, they are at the bottom of most interesting generic code, just as assembly code is at the bottom of most interesting ordinary code. Like assembly code, **requires**-expressions should not be seen in ordinary code. They belong in the implementation of abstractions. If you see **requires requires** in your code, it is probably too low level and will eventually become a problem.

The use of **requires requires** in **advance()** is deliberately inelegant and hackish. Note that I "forgot" to specify += and the required return types for the operations. Therefore, some uses of the version of **advance()** will pass concept checking and still not compile. You have been warned! The proper random-access version of **advance()** is simpler and more readable:

```
template<random_access_iterator Iter>
void advance(Iter p, int n)           // move p n elements forward
{
    p+=n;           // a random-access iterator has +=
}
```

Prefer use of properly named concepts with well-specified semantics (§8.2.4) and primarily use **requires**-expressions in the definition of those.

8.2.4 Definition of Concepts（定义概念）

We find useful concepts, such as **forward_iterator** in libraries, including the standard library (§14.5). As for classes and functions, it is usually easier to use a concept from a good library than to write a new one, but simple concepts are not hard to define. Names from the standard library, such as **random_access_iterator** and **vector**, are in lower case. Here, I use the convention to capitalize the names of concepts I have defined myself, such as **Sequence** and **Vector**.

A concept is a compile-time predicate specifying how one or more types can be used. Consider first one of the simplest examples:

```
template<typename T>
concept Equality_comparable =
    requires (T a, T b) {
        { a == b } -> Boolean;     // compare Ts with ==
        { a != b } -> Boolean;     // compare Ts with !=
    };
```

Equality_comparable is the concept we use to ensure that we can compare values of a type equal and non-equal. We simply say that, given two values of the type, they must be comparable using == and != and the result of those operations must be Boolean. For example:

```
static_assert(Equality_comparable<int>);        // succeeds

struct S { int a; };
static_assert(Equality_comparable<S>);          // fails because structs don't automatically get == and !=
```

The definition of the concept **Equality_comparable** is exactly equivalent to the English description and not any longer. The value of a **concept** is always **bool**.

The result of an { ... } specified after a -> must be a concept. Unfortunately, there isn't a standard-library **boolean** concept, so I defined one (§14.5). **Boolean** simply means a type that can be used as a condition.

Defining **Equality_comparable** to handle nonhomogeneous comparisons is almost as easy:

```
template<typename T, typename T2 =T>
concept Equality_comparable =
    requires (T a, T2 b) {
        { a == b } -> Boolean;     // compare a T to a T2 with ==
        { a != b } -> Boolean;     // compare a T to a T2 with !=
        { b == a } -> Boolean;     // compare a T2 to a T with ==
        { b != a } -> Boolean;     // compare a T2 to a T with !=
    };
```

The **typename T2 =T** says that if we don't specify a second template argument, **T2** will be the same as **T**; **T** is a *default template argument*.

We can test **Equality_comparable** like this:

```
static_assert(Equality_comparable<int,double>);  // succeeds
static_assert(Equality_comparable<int>);          // succeeds (T2 is defaulted to int)
static_assert(Equality_comparable<int,string>);   // fails
```

This **Equally_comparable** is almost identical with the standard-library **equality_comparable** (§14.5).

We can now define a concept that requires arithmetic to be valid between numbers. First we need to define **Number**:

```
template<typename T, typename U = T>
concept Number =
    requires(T x, U y) {  // Something with arithmetic operations and a zero
        x+y; x-y; x*y; x/y;
        x+=y; x-=y; x*=y; x/=y;
        x=x;        // copy
        x=0;
    };
```

This makes no assumptions about the result types, but that's adequate for simple uses. Given one argument type, **Number<X>** checks whether **X** Has the desired properties of a **Number**. Given two arguments, **Number<X,Y>** checks that the two types can be used together with the required operations. From that, we can define our **Arithmetic** concept (§8.2.1):

```
template<typename T, typename U = T>
concept Arithmetic = Number<T,U> && Number<U,T>;
```

For a more complex example, consider a sequence:

```
template<typename S>
concept Sequence = requires (S a) {
     typename range_value_t<S>;                    // S must have a value type
     typename iterator_t<S>;                       // S must have an iterator type

     { a.begin() } -> same_as<iterator_t<S>>;      // S must have a begin() that returns an iterator
     { a.end() } -> same_as<iterator_t<S>>;

     requires input_iterator<iterator_t<S>>;       // S's iterator must be an input_iterator
     requires same_as<range_value_t<S>, iter_value_t<S>>;
};
```

For a type **S** to be a **Sequence**, it must provide a value type (the type of its elements; see §13.1) and an iterator type (the type of its iterators). Here, I used the standard-library associate types **range_value_t<S>** and **iterator_t<S>** (§16.4.4) to express that. It must also ensure that there exist **begin()** and **end()** functions that return **S**'s iterators, as is idiomatic for standard-library containers (§12.3). Finally, **S**'s iterator type must be at least an **input_iterator**, and the value types of the elements and the iterator must be the same.

The hardest concepts to define are the ones that represent fundamental language concepts. Consequently, it is best to use a set from an established library. For a useful collection, see §14.5. In particular, there is a standard-library concept that allows us to bypass the complexity of the definition of **Sequence**:

```
template<typename S>
concept Sequence = input_range<S>;    // simple to write and general
```

Had I restricted my notion of "**S**'s value type to **S::value_type**, I could have used a simple **Value_type**:

```
template<class S>
using Value_type = typename S::value_type;
```

That's a useful technique for expressing simple notions concisely and for hiding complexity. The definition of the standard **value_type_t** is fundamentally similar, but a bit more complicated because it handles sequences that don't have a member called **value_type** (e.g., built-in arrays).

8.2.4.1 Definition Checking（定义时检查）

The concepts specified for a template are used to check arguments at the point of use of the template. They are *not* used to check the use of the parameters in the definition of the template. For example:

```
template<equality_comparable T>
bool cmp(T a, T b)
{
    return a<b;
}
```

Here the concept guarantees the presence of == but not <:

```
bool b0 = cmp(cout,cerr);        // error: ostream doesn't support ==
bool b1 = cmp(2,3);              // OK: returns true
bool b2 = cmp(2+3i,3+4i);        // error: complex<double> doesn't support <
```

The check of concepts catches the attempt to pass the **ostreams**, but accepts the **ints** and the **complex<double>**s because those two types support ==. However, **int** supports < so **cmp(2,3)** compiles, whereas **cmp(2+3i,3+4i)** is rejected when the body of **cmp()** is checked and instantiated for **complex<double>** that does not support <.

Delaying the final check of the template definition until instantiation time gives two benefits:

- We can use incomplete concepts during development. That allows us to gain experience while developing concepts, types, and algorithms, and to gradually improve checking.
- We can insert debug, tracing, telemetry, etc. code into a template without affecting its interface. Changing an interface can cause massive recompilation.

Both are important when developing and maintaining large code bases. The price we pay for that important benefit is that some errors, such as using < where only == is guaranteed, are caught very late in the compilation process (§8.5).

8.2.5 Concepts and auto（概念与 auto）

The keyword **auto** can be used to indicate that an object should have the type of its initializer (§1.4.2):

```
auto x = 1;                          // x is an int
auto z = complex<double>{1,2};       // z is a complex<double>
```

However, initialization doesn't just take place in the simple variable definitions:

```
auto g() { return 99; }              // g() returns an int

int f(auto x) { /* ... */ }          // take an argument of any type

int x = f(1);                        //  this f() takes an int
int z = f(complex<double>{1,2});     // this f() takes a complex<double>
```

The keyword **auto** denotes the least constrained concept for a value: it simply requires that it must be a value of some type. Taking an **auto** parameter makes a function into a function template.

Given concepts, we can strengthen requirements of all such initializations by preceding **auto** by a concept. For example:

```
auto twice(Arithmetic auto x) { return x+x; }  // just for numbers
auto thrice(auto x) { return x+x+x; }          // for anything with a +

auto x1 = twice(7);   // OK: x1==14
string s "Hello ";
auto x2 = twice(s);   // error: a string is not Arithmetic
auto x3 = thrice(s);  // OK x3=="Hello Hello Hello "
```

In addition to their use for constraining function arguments, concepts can constrain the initialization of variables:

```
auto ch1 = open_channel("foo");              // works with whatever open_channel() returns
Arithmetic auto ch2 = open_channel("foo");   // error: a channel is not Arithmetic
Channel auto ch3 = open_channel("foo");      // OK: assuming Channel is an appropriate concept
                                             // and that open_channel() returns one
```

This comes in very handy to counter overuse of **auto** and to document requirements on code using generic functions.

For readability and debugging it is often important that a type error is caught as close to its origin as possible. Constraining a return type can help:

```
Number auto some_function(int x)
{
    // ...
    return fct(x);    // an error unless fct(x) returns a Number
    // ...
}
```

Naturally, we could have achieved that by introducing a local variable:

```
auto some_function(int x)
{
    // ...
    Number auto y = fct(x);    // an error unless fct(x) returns a Number
    return y;
    // ...
}
```

However, that's a bit verbose and not all types can be cheaply copied.

8.2.6 Concepts and Types（类型与概念）

A type

- Specifies the set of operations that can be applied to an object, implicitly and explicitly
- Relies on function declarations and language rules
- Specifies how an object is laid out in memory

A single-argument concept

- Specifies the set of operations that can be applied to an object, implicitly and explicitly
- Relies on use patterns reflecting function declarations and language rules
- Says nothing about the layout of the object

- Enables the use of a set of types

Thus, constraining code with concepts gives more flexibility than constraining with types. In addition, concepts can define the relationship among several arguments. My ideal is that eventually most functions will be defined as template functions with their arguments constrained by concepts. Unfortunately, the notational support for that is not yet perfect: we have to use a concept as an adjective, rather that a noun. For example:

```
void sort(Sortable auto&);      // 'auto' required
void sort(Sortable&);           // error: 'auto' required after concept name
```

8.3 Generic Programming（泛型编程）

The form of *generic programming* directly supported by C++ centers around the idea of abstracting from concrete, efficient algorithms to obtain generic algorithms that can be combined with different data representations to produce a wide variety of useful software [Stepanov,2009]. The abstractions that represent the fundamental operations and data structures are called *concepts*.

8.3.1 Use of Concepts（概念的使用）

Good, useful concepts are fundamental and are discovered more than they are designed. Examples are integer and floating-point number (as defined even in Classic C [Kernighan,1978]), sequence, and more general mathematical concepts, such as ring and vector space. They represent the fundamental concepts of a field of application. That is why they are called "concepts." Identifying and formalizing concepts to the degree necessary for effective generic programming can be a challenge.

For basic use, consider the concept regular (§14.5). A type is regular when it behaves much like an int or a vector. An object of a regular type

- can be default constructed.
- can be copied (with the usual semantics of copy, yielding two objects that are independent and compare equal) using a constructor or an assignment.
- can be compared using == and !=.
- doesn't suffer technical problems from overly clever programming tricks.

A string is another example of a regular type. Like int, string is also totally_ordered (§14.5). That is, two strings can be compared using <, <=, >, >=, and <=> with the appropriate semantics.

A concept is not just a syntactic notion, it is fundamentally about semantics. For example, don't define + to divide; that would not match the requirements for any reasonable number. Unfortunately, we do not yet have any language support for expressing semantics, so we have to rely on expert knowledge and common sense to get semantically meaningful concepts. Do not define semantically meaningless concepts, such as Addable and Subtractable. Instead, rely on domain knowledge to define concepts that match fundamental concepts in an application domain.

8.3.2 Abstraction Using Templates（使用模板实现抽象）

Good abstractions are carefully grown from concrete examples. It is not a good idea to try to "abstract" by trying to prepare for every conceivable need and technique; in that direction lies inelegance and code bloat. Instead, start with one – and preferably more – concrete examples from real use and try to eliminate inessential details. Consider:

```
double sum(const vector<int>& v)
{
      double res = 0;
      for (auto x : v)
            res += x;
      return res;
}
```

This is obviously one of many ways to compute the sum of a sequence of numbers.

Consider what makes this code less general than it needs to be:

- Why just ints?
- Why just vectors?
- Why accumulate in a double?
- Why start at 0?
- Why add?

Answering the first four questions by making the concrete types into template arguments, we get the simplest form of the standard-library accumulate algorithm:

```
template<forward_iterator Iter, Arithmetic<iter_value_t<Iter>> Val>
Val accumulate(Iter first, Iter last, Val res)
{
      for (auto p = first; p!=last; ++p)
            res += *p;
      return res;
}
```

Here, we have:

- The data structure to be traversed has been abstracted into a pair of iterators representing a sequence (§8.2.4, §13.1).
- The type of the accumulator has been made into a parameter.
- The type of the accumulator must be arithmetic .
- The type of the accumulator must work with the iterator's value type (the element type of the sequence).
- The initial value is now an input; the type of the accumulator is the type of this initial value.

A quick examination or – even better – measurement will show that the code generated for calls with a variety of data structures is identical to what you get from hand-coded examples. Consider:

```
void use(const vector<int>& vec, const list<double>& lst)
{
      auto sum = accumulate(begin(vec),end(vec),0.0);   // accumulate in a double
      auto sum2 = accumulate(begin(lst),end(lst),sum);
      // ...
}
```

The process of generalizing from a concrete piece of code (and preferably from several) while preserving performance is called *lifting*. Conversely, the best way to develop a template is often to

- first, write a concrete version
- then, debug, test, and measure it
- finally, replace the concrete types with template arguments.

Naturally, the repetition of **begin()** and **end()** is tedious, so we can simplify the user interface a bit:

```
template<forward_range R, Arithmetic<value_type_t<R>> Val>
Val accumulate(const R& r, Val res = 0)
{
	for (auto x : r)
		res += x;
	return res;
}
```

A *range* is a standard-library concept representing a sequence with **begin()** and **end()** (§13.1). For full generality, we can abstract the **+=** operation also; see §17.3.

Both the pair-of-iterators and the range version of **accumulate()** are useful: the pair-of-iterators version for generality, the range version for simplicity of common uses.

8.4 Variadic Templates（可变参数模板）

A template can be defined to accept an arbitrary number of arguments of arbitrary types. Such a template is called a *variadic template*. Consider a simple function to write out values of any type that has a << operator:

```
void user()
{
	print("first: ", 1, 2.2, "hello\n"s);                    // first:  1 2.2 hello

	print("\nsecond: ", 0.2, 'c', "yuck!"s, 0, 1, 2, '\n');  // second: 0.2 c yuck! 0 1 2
}
```

Traditionally, implementing a variadic template has been to separate the first argument from the rest and then recursively call the variadic template for the tail of the arguments:

```
template<typename T>
concept Printable = requires(T t) { std::cout << t; } // just one operation!

void print()
{
	// what we do for no arguments: nothing
}

template<Printable T, Printable... Tail>
void print(T head, Tail... tail)
{
	cout << head << ' ';      // first, what we do for the head
	print(tail...);           // then, what we do for the tail
}
```

The **Printable...** indicates that **Tail** is a sequence of types. The **Tail...** indicates that **tail** is a sequence of values of the types in Tail. A parameter declared with a ... is called a *parameter pack*. Here, tail is a (function argument) parameter pack where the elements are of the types found in the (template argument) parameter pack Tail. So, print() can take any number of arguments of any types.

A call of print() separates the arguments into a head (the first) and a tail (the rest). The head is printed and then print() is called for the tail. Eventually, of course, tail will become empty, so we need the no-argument version of print() to deal with that. If we don't want to allow the zero-argument case, we can eliminate that print() using a compile-time if:

```
template<Printable T, Printable... Tail>
void print(T head, Tail... tail)
{
        cout << head << ' ';
        if constexpr(sizeof...(tail)> 0)
              print(tail...);
}
```

I used a compile-time if (§7.4.3), rather than a plain run-time if, to avoid a final call print() from being generated. Given that, the "empty" print() need not be defined.

The strength of variadic templates is that they can accept any arguments you care to give them. Weaknesses include

- The recursive implementations can be tricky to get right.
- The type checking of the interface is a possibly elaborate template program.
- The type checking code is ad hoc, rather than defined in the standard.
- The recursive implementations can be surprisingly expensive in compile time and compiler memory requirements.

Because of their flexibility, variadic templates are widely used in the standard library, and occasionally wildly overused.

8.4.1 Fold Expressions（折叠表达式）

To simplify the implementation of simple variadic templates, C++ offers a limited form of iteration over elements of a parameter pack. For example:

```
template<Number... T>
int sum(T... v)
{
    return (v + ... + 0);        // add all elements of v starting with 0
}
```

This sum() can take any number of arguments of any types:

```
int x = sum(1, 2, 3, 4, 5); // x becomes 15
int y = sum('a', 2.4, x);     // y becomes 114 (2.4 is truncated and the value of 'a' is 97)
```

The body of sum uses a fold expression:

```
return (v + ... + 0);   // add all elements of v to 0
```

Here, (v+...+0) means add all the elements of v starting with the initial value 0. The first element to be added is the "rightmost" (the one with the highest index): (v[0]+(v[1]+(v[2]+(v[3]+(v[4]+0))))). That is, starting from the right where the 0 is. It is called a *right fold*. Alternatively, we could have used a *left fold*:

```
template<Number... T>
int sum2(T... v)
{
    return (0 + ... + v); // add all elements of v to 0
}
```

Now, the first element to be added is the "leftmost" (the one with the lowest index): (((((0+v[0])+v[1])+v[2])+v[3])+v[4]). That is, starting from the left where the 0 is.

Fold is a very powerful abstraction, clearly related to the standard-library accumulate(), with a variety of names in different languages and communities. In C++, the fold expressions are currently restricted to simplify the implementation of variadic templates. A fold does not have to perform numeric computations. Consider a famous example:

```
template<Printable ...T>
void print(T&&... args)
{
    (std::cout << ... << args) << '\n';    // print all arguments
}

print("Hello!"s,' ',"World ",2017);    // (((((std::cout << "Hello!"s) << ' ') << "World ") << 2017) << '\n');
```

Why 2017? Because fold() was added to C++ in 2017 (§19.2.3).

8.4.2 Forwarding Arguments（完美转发参数）

Passing arguments unchanged through an interface is an important use of variadic templates. Consider a notion of a network input channel for which the actual method of moving values is a parameter. Different transport mechanisms have different sets of constructor parameters:

```
template<concepts::InputTransport Transport>
class InputChannel {
public:
        // ...
        InputChannel(Transport::Args&&... transportArgs)
                : _transport(std::forward<TransportArgs>(transportArgs)...)
    {}
        // ...
        Transport _transport;
};
```

The standard-library function forward() (§16.6) is used to move the arguments unchanged from the InputChannel constructor to the Transport constructor.

The point here is that the writer of InputChannel can construct an object of type Transport without having to know what arguments are required to construct a particular Transport. The implementer of InputChannel needs only to know the common user interface for all Transport objects.

Forwarding is very common in foundational libraries where generality and low run-time overhead are necessary and very general interfaces are common.

8.5 Template Compilation Model （模板编译模型）

At the point of use, the arguments for a template are checked against its concepts. Errors found here will be reported immediately. What cannot be checked at this point, such as arguments for unconstrained template parameters, is postponed until code is generated for the template with a set of template arguments: "at template instantiation time."

An unfortunate side effect of instantiation-time type checking is that a type error can be detected uncomfortably late (§8.2.4.1). Also, late checking often results in spectacularly bad error messages because the compiler does not have type information giving hints to the programmer's intent and often detects a problem only after combining information from several places in the program.

The instantiation-time type checking provided for templates checks the use of arguments in the template definition. This provides a compile-time variant of what is often called *duck typing* ("If it walks like a duck and it quacks like a duck, it's a duck"). Or – using more technical terminology – we operate on values, and the presence and meaning of an operation depend solely on its operand values. This differs from the alternative view that objects have types, which determine the presence and meaning of operations. Values "live" in objects. This is the way objects (e.g., variables) work in C++, and only values that meet an object's requirements can be put into it. What is done at compile time using templates mostly does not involve objects, only values. The exception is local variables in a **constexpr** function (§1.6) that are used as objects inside the compiler.

To use an unconstrained template, its definition (not just its declaration) must be in scope at its point of use. When using header files and **#include**, this means that template definitions are found in header files, rather than **.cpp** files. For example, the standard header **<vector>** holds the definition of **vector**.

This changes when we start to use modules (§3.2.2). Using modules, the source code can be organized in the same way for ordinary functions and template functions. A module is semi-compiled into a representation that makes it fast to **import** and use. Think of that representation as an easily traversed graph containing all available scope and type information and supported by a symbol table allowing quick access to individual entities.

8.6 Advice（建议）

[1] Templates provide a general mechanism for compile-time programming; §8.1.
[2] When designing a template, carefully consider the concepts (requirements) assumed for its template arguments; §8.3.2.
[3] When designing a template, use a concrete version for initial implementation, debugging, and measurement; §8.3.2.
[4] Use concepts as a design tool; §8.2.1.
[5] Specify concepts for all template arguments; §8.2; [CG: T.10].
[6] Whenever possible use named concepts (e.g., standard-library concepts); §8.2.4, §14.5; [CG: T.11].
[7] Use a lambda if you need a simple function object in one place only; §7.3.2.
[8] Use templates to express containers and ranges; §8.3.2; [CG: T.3].

[9] Avoid "concepts" without meaningful semantics; §8.2; [CG: T.20].
[10] Require a complete set of operations for a concept; §8.2; [CG: T.21].
[11] Use named concepts §8.2.3.
[12] Avoid **requires requires**; §8.2.3.
[13] **auto** is the least constrained concept §8.2.5.
[14] Use variadic templates when you need a function that takes a variable number of arguments of a variety of types; §8.4.
[15] Templates offer compile-time "duck typing"; §8.5.
[16] When using header files, **#include** template definitions (not just declarations) in every translation unit that uses them; §8.5.
[17] To use a template, make sure its definition (not just its declaration) is in scope; §8.5.
[18] Unconstrained templates offer compile-time "duck typing"; §8.5.

9

Library Overview

（标准库）

Why waste time learning
when ignorance is instantaneous?
– Hobbes

9.1 Introduction （引言）

No significant program is written in just a bare programming language. First, a set of libraries is developed. These then form the basis for further work. Most programs are tedious to write in the bare language, whereas just about any task can be rendered simple by the use of good libraries.

Continuing from Chapters 1–8, Chapters 9–18 give a quick tour of key standard-library facilities. I very briefly present useful standard-library types, such as **string**, **ostream**, **variant**, **vector**, **map**, **path**, **unique_ptr**, **thread**, **regex**, **system_clock**, **time_zone**, and **complex**, as well as the most common ways of using them.

As in Chapters 1–8, you are strongly encouraged not to be distracted or discouraged by an incomplete understanding of details. The purpose of this chapter is to convey a basic understanding of the most useful library facilities.

The specification of the standard library is over two thirds of the ISO C++ standard. Explore it, and prefer it to home-made alternatives. Much thought has gone into its design, more still into its implementations, and much effort will go into its maintenance and extension.

The standard-library facilities described in this book are part of every complete C++ implementation. In addition to the standard-library components, most implementations offer "graphical user interface" systems (GUIs), Web interfaces, database interfaces, etc. Similarly, most application-development environments provide "foundation libraries" for corporate or industrial "standard" development and/or execution environments. Beyond that, there are many thousands of libraries supporting specialized application areas. Here, I do not describe libraries, systems, or environments beyond the standard-libraries. The intent is to provide a self-contained description of C++ as defined by its standard [C++,2020] and to keep the examples portable. Naturally, a programmer is encouraged to explore the more extensive facilities available on most systems.

9.2 Standard-Library Components（标准库组件）

The facilities provided by the standard library can be classified like this:

- Run-time language support (e.g., for allocation, exceptions, and run-time type information).
- The C standard library (with very minor modifications to minimize type system violations).
- Strings with support for international character sets, localization, and read-only views of substrings (§10.2).
- Support for regular expression matching (§10.4).
- I/O streams is an extensible framework for input and output to which users can add their own types, streams, buffering strategies, locales, and character sets (Chapter 11). It also offers facilities for flexible output formatting (§11.6.2).
- A library for manipulating file systems in a portable manner (§11.9).
- A framework of containers (such as **vector** and **map**; Chapter 12) and algorithms (such as **find()**, **sort()**, and **merge()**; Chapter 13). This framework, conventionally called the STL [Stepanov,1994], is extensible so users can add their own containers and algorithms.
- Ranges (§14.1), including views (§14.2), generators (§14.3), and pipes (§14.4).
- Concepts for fundamental types and ranges (§14.5).
- Support for numerical computation, such as standard mathematical functions, complex numbers, vectors with arithmetic operations, mathematical constants, and random-number generators (§5.2.1 and Chapter 16).
- Support for concurrent programming, including **thread**s and locks (Chapter 18). The concurrency support is foundational so that users can add support for new models of concurrency as libraries.
- Synchronous and asynchronous coroutines (§18.6).
- Parallel versions of most STL algorithms and of some numerical algorithms, such as **sort()** (§13.6) and **reduce()** (§17.3.1).
- Utilities to support metaprogramming (e.g., type functions; §16.4), STL-style generic programming (e.g., **pair**; §15.3.3), and general programming (e.g., **variant** and **optional**; §15.4.1, §15.4.2).
- "Smart pointers" for resource management (e.g., **unique_ptr** and **shared_ptr**; §15.2.1).
- Special-purpose containers, such as **array** (§15.3.1), **bitset** (§15.3.2), and **tuple** (§15.3.3).
- Support for absolute time and durations, e.g., **time_point** and **system_clock** (§16.2.1).
- Support for calendars, e.g., **month** and **time_zone** (§16.2.2, §16.2.3).
- Suffixes for popular units, such as **ms** for milliseconds and **i** for imaginary (§6.6).
- Ways of manipulating sequences of elements, such as views (§14.2), **string_view**s (§10.3), and **span**s (§15.2.2).

The main criteria for including a class in the library were that:

- It could be helpful to almost every C++ programmer (both novices and experts).
- It could be provided in a general form that did not add significant overhead compared to a simpler version of the same facility.
- Simple uses should be easy to learn (relative to the inherent complexity of their task).

Essentially, the C++ standard library provides the most common fundamental data structures together with the fundamental algorithms used on them.

9.3 Standard-Library Organization （标准库的组织）

The facilities of the standard library are placed in namespace **std** and made available to users through modules or headed files.

9.3.1 Namespaces （命名空间）

Every standard-library facility is provided through some standard header. For example:

```
#include<string>
#include<list>
```

This makes the standard **string** and **list** available.

The standard library is defined in a namespace (§3.3) called **std**. To use standard-library facilities, the **std::** prefix can be used:

```
std::string sheep {"Four legs Good; two legs Baaad!"};
std::list<std::string> slogans {"War is Peace", "Freedom is Slavery", "Ignorance is Strength"};
```

For brevity, I rarely use the **std::** prefix in examples. Neither do I **#include** or **import** the necessary headers or modules explicitly. To compile and run the program fragments here, you must make the relevant parts of the standard library available. For example:

```
#include<string>                // make the standard string facilities accessible
using namespace std;             // make std names available without std:: prefix

string s {"C++ is a general-purpose programming language"};     // OK: string is std::string
```

It is generally in poor taste to dump every name from a namespace into the global namespace. However, in this book, I use the standard library exclusively and it is good to know what it offers.

The standard library offers several sub-namespaces to **std** that can be accessed only through an explicit action:

- **std::chrono**: all facilities from chrono, including **std::literals::chrono_literals** (§16.2).
- **std::literals::chrono_literals**: suffixes **y** for years, **d** for days, **h** for hours, **min** for minutes, **ms** for milliseconds, **ns** for nanoseconds, **s** for seconds, and **us** for microseconds (§16.2).
- **std::literals::complex_literals**: suffixes **i** for imaginary doubles, **if** for imaginary floats, and **il** for imaginary long doubles (§6.6).
- **std::literals::string_literals**: suffix **s** for strings (§6.6, §10.2).
- **std::literals::string_view_literals**: suffix **sv** for string views (§10.3).

- **std::numbers** for mathematical constants (§17.9).
- **std::pmr** for polymorphic memory resources (§12.7).

To use a suffix from a sub-namespace, we have to introduce it into the namespace in which we want to use it. For example:

```
// no mention of complex_literals
auto z1 = 2+3i;         // error: no suffix 'i'

using namespace literals::complex_literals; // make the complex literals visible
auto z2 = 2+3i;         // ok: z2 is a complex<double>
```

There is no coherent philosophy for what should be in a sub-namespace. However, suffixes cannot be explicitly qualified so we can only bring in a single set of suffixes into a scope without risking ambiguities. Therefore suffixes for a library meant to work with other libraries (that might define their own suffixes) are placed in sub-namespaces.

9.3.2 The ranges namespace（ranges 命名空间）

The standard-library offers algorims, such as **sort()** and **copy()**, in two versions:

- A traditional sequence version taking a pair of iterators; e.g., **sort(begin(v),v.end())**
- A range version taking a single range; e.g., **sort(v)**

Ideally, these two versions should overload perfectly without any special effort. However, they don't. For example:

```
using namespace std;
using namespace ranges;

void f(vector<int>& v)
{
        sort(v.begin(),v.end());     // error:ambiguous
        sort(v);                     // error: ambiguous
}
```

To protect against ambiguities when using traditional unconstrained templates, the standard requires that we explicitly introduce the range version of a standard-library algorithm into a scope:

```
using namespace std;

void g(vector<int>& v)
{
        sort(v.begin(),v.end());     // OK
        sort(v);                     // error: no matching function (in std)
        ranges::sort(v);             // OK
        using ranges::sort;          // sort(v) OK from here on
        sort(v);                     // OK
}
```

9.3.3 Modules（模块）

There are not yet any standard-library modules. C++23 is likely to remedy this omission (caused by lack of committee time). For now, I use module std that is likely to become standard, offering all facilities from namespace std. See Appendix A.

9.3.4 Headers（头文件）

Here is a selection of standard-library headers, all supplying declarations in namespace std:

Selected Standard Library Headers		
<algorithm>	copy(), find(), sort()	Chapter 13
<array>	array	§15.3.1
<chrono>	duration, time_point, month, time_zone	§16.2
<cmath>	sqrt(), pow()	§17.2
<complex>	complex, sqrt(), pow()	§17.4
<concepts>	floating_point, copyable, predicate, invocable	§14.5
<filesystem>	path	§11.9
<format>	format()	§11.6.2
<fstream>	fstream, ifstream, ofstream	§11.7.2
<functional>	function, greater_equal, hash, range_value_t	Chapter 16
<future>	future, promise	§18.5
<ios>	hex, dec, scientific, fixed, defaultfloat	§11.6.2
<iostream>	istream, ostream, cin, cout	Chapter 11
<map>	map, multimap	§12.6
<memory>	unique_ptr, shared_ptr, allocator	§15.2.1
<random>	default_random_engine, normal_distribution	§17.5
<ranges>	sized_range, subrange, take(), split(), iterator_t	§14.1
<regex>	regex, smatch	§10.4
<string>	string, basic_string	§10.2
<string_view>	string_view	§10.3
<set>	set, multiset	§12.8
<sstream>	istringstream, ostringstream	§11.7.3
<stdexcept>	length_error, out_of_range, runtime_error	§4.2
<tuple>	tuple, get<>(), tuple_size<>	§15.3.4
<thread>	thread	§18.2
<unordered_map>	unordered_map, unordered_multimap	§12.6
<utility>	move(), swap(), pair	Chapter 16
<variant>	variant	§15.4.1
<vector>	vector	§12.2

This listing is far from complete.

Headers from the C standard library, such as <stdlib.h> are provided. For each such header there is also a version with its name prefixed by c and the .h removed. This version, such as <cstdlib> places its declarations in both the std and global namespace.

The headers reflect the history of the development of the standard library. Consequently, they are not always as logical and easy to remember as we would like. That's one reason to use a module, such as **std** (§9.3.3), instead.

9.4 Advice（建议）

[1] Don't reinvent the wheel; use libraries; §9.1; [CG: SL.1.]
[2] When you have a choice, prefer the standard library over other libraries; §9.1; [CG: SL.2].
[3] Do not think that the standard library is ideal for everything; §9.1.
[4] If you don't use modules, remember to **#include** the appropriate headers; §9.3.1.
[5] Remember that standard-library facilities are defined in namespace **std**; §9.3.1; [CG: SL.3].
[6] When using **ranges**, remember to explicitly qualify algorithm names; §9.3.2.
[7] Prefer **import**ing **module**s over **#include**ing header files (§9.3.3).

10

Strings and Regular Expressions
（字符串和正则表达式）

Prefer the standard to the offbeat.
– Strunk & White

10.1 Introduction（引言）

Text manipulation is a major part of most programs. The C++ standard library offers a **string** type to save most users from C-style manipulation of arrays of characters through pointers. A **string_view** type allows us to manipulate sequences of characters however they may be stored (e.g., in a **std::string** or a **char[]**). In addition, regular expression matching is offered to help find patterns in text. The regular expressions are provided in a form similar to what is common in most modern languages. Both **string**s and **regex** objects can use a variety of character types (e.g., Unicode).

10.2 Strings（字符串）

The standard library provides a **string** type to complement the string literals (§1.2.1); **string** is a **regular** type (§8.2, §14.5) for owning and manipulating a sequence of characters of various character types. The **string** type provides a variety of useful string operations, such as concatenation. For example:

```
string compose(const string& name, const string& domain)
{
    return name + '@' + domain;
}

auto addr = compose("dmr","bell-labs.com");
```

Here, addr is initialized to the character sequence dmr@bell-labs.com. "Addition" of strings means concatenation. You can concatenate a string, a string literal, a C-style string, or a character to a string. The standard string has a move constructor, so returning even long strings by value is efficient (§6.2.2).

In many applications, the most common form of concatenation is adding something to the end of a string. This is directly supported by the += operation. For example:

```
void m2(string& s1, string& s2)
{
    s1 = s1 + '\n';  // append newline
    s2 += '\n';      // append newline
}
```

The two ways of adding to the end of a string are semantically equivalent, but I prefer the latter because it is more explicit about what it does, more concise, and possibly more efficient.

A string is mutable. In addition to = and +=, subscripting (using [ ]) and substring operations are supported. For example:

```
string name = "Niels Stroustrup";

void m3()
{
    string s = name.substr(6,10);       // s = "Stroustrup"
    name.replace(0,5,"nicholas");       // name becomes "nicholas Stroustrup"
    name[0] = toupper(name[0]);         // name becomes "Nicholas Stroustrup"
}
```

The substr() operation returns a string that is a copy of the substring indicated by its arguments. The first argument is an index into the string (a position), and the second is the length of the desired substring. Since indexing starts from 0, s gets the value Stroustrup.

The replace() operation replaces a substring with a value. In this case, the substring starting at 0 with length 5 is Niels; it is replaced by nicholas. Finally, I replace the initial character with its uppercase equivalent. Thus, the final value of name is Nicholas Stroustrup. Note that the replacement string need not be the same size as the substring that it is replacing.

Among the many useful string operations are assignment (using =), subscripting (using [ ] or at() as for vector; §12.2.2), comparison (using == and !=), and lexicographical ordering (using <, <=, >, and >=), iteration (using iterators, begin(), and end() as for vector; §13.2), input (§11.3), and streaming (§11.7.3).

Naturally, strings can be compared against each other, against C-style strings §1.7.1), and against string literals. For example:

```
string incantation;

void respond(const string& answer)
{
      if (answer == incantation) {
            // ... perform magic ...
      }
      else if (answer == "yes") {
            // ...
      }
      // ...
}
```

If you need a C-style string (a zero-terminated array of **char**), **string** offers read-only access to its contained characters (**c_str()** and **data()**). For example:

```
void print(const string& s)
{
      printf("For people who like printf: %s\n",s.c_str());        // s.c_str() returns a pointer to s' characters
      cout << "For people who like streams: " << s << '\n';
}
```

A string literal is by definition a **const char***. To get a literal of type **std::string** use an **s** suffix. For example:

```
auto cat = "Cat"s;    // a std::string
auto dog = "Dog";     // a C-style string: a const char*
```

To use the **s** suffix, you need to use the namespace **std::literals::string_literals** (§6.6).

10.2.1 string Implementation（string 的实现）

Implementing a string class is a popular and useful exercise. However, for general-purpose use, our carefully crafted first attempts rarely match the standard **string** in convenience or performance. These days, **string** is usually implemented using the *short-string optimization*. That is, short string values are kept in the **string** object itself and only longer strings are placed on free store. Consider:

```
string s1 {"Annemarie"};                // short string
string s2 {"Annemarie Stroustrup"};     // long string
```

The memory layout will be something like this:

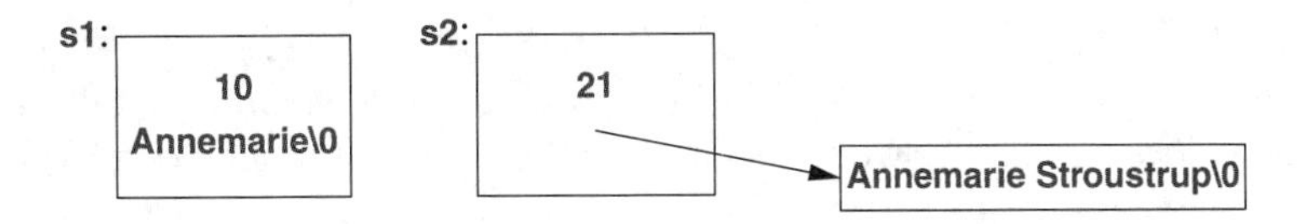

When a **string**'s value changes from a short to a long string (and vice versa) its representation adjusts appropriately. How many characters can a "short" string have? That's implementation defined, but "about 14 characters" isn't a bad guess.

The actual performance of **strings** can depend critically on the run-time environment. In particular, in multi-threaded implementations, memory allocation can be relatively costly. Also, when lots of strings of differing lengths are used, memory fragmentation can result. These are the main reasons that the short-string optimization has become ubiquitous.

To handle multiple character sets, **string** is really an alias for a general template **basic_string** with the character type **char**:

```
template<typename Char>
class basic_string {
    // ... string of Char ...
};

using string = basic_string<char>;
```

A user can define strings of arbitrary character types. For example, assuming we have a Japanese character type **Jchar**, we can write:

```
using Jstring = basic_string<Jchar>;
```

Now we can do all the usual string operations on **Jstring**, a string of Japanese characters.

10.3 String Views （字符串视图）

The most common use of a sequence of characters is to pass it to some function to read. This can be achieved by passing a **string** by value, a reference to a string, or a C-style string. In many systems there are further alternatives, such as string types not offered by the standard. In all of these cases, there are extra complexities when we want to pass a substring. To address this, the standard library offers **string_view**; a **string_view** is basically a (pointer,length) pair denoting a sequence of characters:

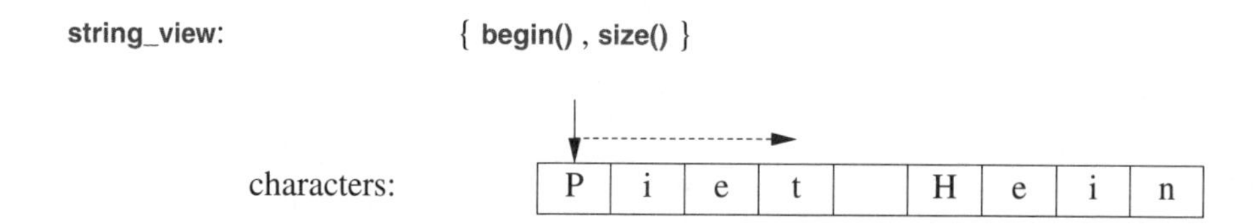

A **string_view** gives access to a contiguous sequence of characters. The characters can be stored in many possible ways, including in a **string** and in a C-style string. A **string_view** is like a pointer or a reference in that it does not own the characters it points to. In that, it resembles an STL pair of iterators (§13.3).

Consider a simple function concatenating two strings:

```
string cat(string_view sv1, string_view sv2)
{
    string res {sv1};        // initialize from sv1
    return res += sv2;       // append from sv2 and return

}
```

We can call this cat():

```
string king = "Harold";
auto s1 = cat(king,"William");                  // HaroldWilliam: string and const char*
auto s2 = cat(king,king);                       // HaroldHarold: string and string
auto s3 = cat("Edward","Stephen"sv);            // EdwardStephen: const char * and string_view
auto s4 = cat("Canute"sv,king);                 // CanuteHarold
auto s5 = cat({&king[0],2},"Henry"sv);          // HaHenry
auto s6 = cat({&king[0],2},{&king[2],4});       // Harold
```

This cat() has three advantages over the compose() that takes const string& arguments (§10.2):

- It can be used for character sequences managed in many different ways.
- We can easily pass a substring.
- We don't have to create a string to pass a C-style string argument.

Note the use of the sv (''string view'') suffix. To use that, we need to make it visible:

```
using namespace std::literals::string_view_literals;      // §6.6
```

Why bother with a suffix? The reason is that when we pass "Edward" we need to construct a string_view from a const char* and that requires counting the characters. For "Stephen"sv the length is computed at compile time.

A string_view defines a range, so we can traverse its characters. For example:

```
void print_lower(string_view sv1)
{
     for (char ch : sv1)
          cout << tolower(ch);
}
```

One significant restriction of string_view is that it is a read-only view of its characters. For example, you cannot use a string_view to pass characters to a function that modifies its argument to lowercase. For that, you might consider using a span (§15.2.2).

Think of string_view as a kind of pointer; to be used, it must point to something:

```
string_view bad()
{
     string s = "Once upon a time";
     return {&s[5],4};          // bad: returning a pointer to a local
}
```

Here, the returned string will be destroyed before we can use its characters.

The behavior of out-of-range access to a string_view is undefined. If you want guaranteed range checking, use at(), which throws out_of_range for attempted out-of-range access, or gsl::string_span (§15.2.2).

10.4 Regular Expressions（正则表达式）

Regular expressions are a powerful tool for text processing. They provide a way to simply and tersely describe patterns in text (e.g., a U.S. postal code such as **TX 77845**, or an ISO-style date, such as **2009-06-07**) and to efficiently find such patterns. In **<regex>**, the standard library provides support for regular expressions in the form of the **std::regex** class and its supporting functions. To give a taste of the style of the **regex** library, let us define and print a pattern:

```
regex pat {R"(\w{2}\s*\d{5}(-\d{4})?)"};   // U.S. postal code pattern: XXddddd-dddd and variants
```

People who have used regular expressions in just about any language will find **\w{2}\s*\d{5}(-\d{4})?** familiar. It specifies a pattern starting with two letters **\w{2}** optionally followed by some space **\s*** followed by five digits **\d{5}** and optionally followed by a dash and four digits **-\d{4}**. If you are not familiar with regular expressions, this may be a good time to learn about them ([Stroustrup,2009], [Maddock,2009], [Friedl,1997]).

To express the pattern, I use a *raw string literal* starting with **R"(** and terminated by **)"**. This allows backslashes and quotes to be used directly in the string. Raw strings are particularly suitable for regular expressions because they tend to contain a lot of backslashes. Had I used a conventional string, the pattern definition would have been:

```
regex pat {"\\w{2}\\s*\\d{5}(-\\d{4})?"};   // U.S. postal code pattern
```

In **<regex>**, the standard library provides support for regular expressions:

- **regex_match()**: Match a regular expression against a string (of known size) (§10.4.2).
- **regex_search()**: Search for a string that matches a regular expression in an (arbitrarily long) stream of data (§10.4.1).
- **regex_replace()**: Search for strings that match a regular expression in an (arbitrarily long) stream of data and replace them.
- **regex_iterator**: Iterate over matches and submatches (§10.4.3).
- **regex_token_iterator**: Iterate over non-matches.

10.4.1 Searching（搜索）

The simplest way of using a pattern is to search for it in a stream:

```
int lineno = 0;
for (string line; getline(cin,line); ) {          // read into line buffer
      ++lineno;
      smatch matches;                               // matched strings go here
      if (regex_search(line,matches,pat))           // search for pat in line
            cout << lineno << ": " << matches[0] << '\n';
}
```

The **regex_search(line,matches,pat)** searches the **line** for anything that matches the regular expression stored in **pat** and if it finds any matches, it stores them in **matches**. If no match was found, **regex_search(line,matches,pat)** returns **false**. The **matches** variable is of type **smatch**. The "s" stands for "sub" or "string," and an **smatch** is a **vector** of submatches of type **string**. The first element, here **matches[0]**, is the complete match. The result of a **regex_search()** is a collection of matches, typically represented as an **smatch**:

```
void use()
{
      ifstream in("file.txt");        // input file
      if (!in) {                      // check that the file was opened
            cerr << "no file\n";
            return;
      }

      regex pat {R"(\w{2}\s*\d{5}(-\d{4})?)"};    // U.S. postal code pattern

      int lineno = 0;
      for (string line; getline(in,line); ) {
            ++lineno;
            smatch matches;       // matched strings go here
            if (regex_search(line, matches, pat)) {
                  cout << lineno << ": " << matches[0] << '\n';       // the complete match
                  if (1<matches.size() && matches[1].matched)         // if there is a sub-pattern
                                                                      // and if it is matched
                        cout << "\t: " << matches[1] << '\n';         // submatch
            }
      }
}
```

This function reads a file looking for U.S. postal codes, such as **TX77845** and **DC 20500-0001**. An **smatch** type is a container of regex results. Here, **matches[0]** is the whole pattern and **matches[1]** is the optional four-digit subpattern **(-\d{4})?**.

The newline character, **\n**, can be part of a pattern, so we can search for multiline patterns. Obviously, we shouldn't read one line at a time if we want to do that.

The regular expression syntax and semantics are designed so that regular expressions can be compiled into state machines for efficient execution [Cox,2007]. The **regex** type performs this compilation at run time.

10.4.2 Regular Expression Notation （正则表达式的符号表示）

The **regex** library can recognize several variants of the notation for regular expressions. Here, I use the default notation, a variant of the ECMA standard used for ECMAScript (more commonly known as JavaScript). The syntax of regular expressions is based on *special characters:*

Regular Expression Special Characters			
.	Any single character (a "wildcard")	\	Next character has a special meaning
[	Begin character class	*	Zero or more (suffix operation)
]	End character class	+	One or more (suffix operation)
{	Begin count	?	Optional (zero or one) (suffix operation)
}	End count	|	Alternative (or)
(	Begin grouping	^	Start of line; negation
)	End grouping	$	End of line

For example, we can specify a line starting with zero or more **A**s followed by one or more **B**s followed by an optional **C** like this:

```
^A*B+C?$
```

Examples that match:

```
AAAAAAAAAAAABBBBBBBBBC
BC
B
```

Examples that do not match:

```
AAAAA          // no B
  AAAABC       // initial space
AABBCC         // too many Cs
```

A part of a pattern is considered a subpattern (which can be extracted separately from an **smatch**) if it is enclosed in parentheses. For example:

```
\d+-\d+         // no subpatterns
\d+(-\d+)       // one subpattern
(\d+)(-\d+)     // two subpatterns
```

A pattern can be optional or repeated (the default is exactly once) by adding a suffix:

Repetition	
{n}	Exactly n times
{n,}	n or more times
{n,m}	At least n and at most m times
*	Zero or more, that is, {0,}
+	One or more, that is, {1,}
?	Optional (zero or one), that is {0,1}

For example:

```
A{3}B{2,4}C*
```

Examples that match:

```
AAABBC
AAABBB
```

Examples that do not match:

```
AABBC           // too few As
AAABC           // too few Bs
AAABBBBBCCC     // too many Bs
```

A suffix **?** after any of the repetition notations (**?**, *, **+**, and **{ }**) makes the pattern matcher "lazy" or "non-greedy." That is, when looking for a pattern, it will look for the shortest match rather than the longest. By default, the pattern matcher always looks for the longest match; this is known as the *Max Munch rule*. Consider:

```
ababab
```

The pattern (ab)+ matches all of ababab. However, (ab)+? matches only the first ab.

The most common character classifications have names:

Character Classes	
alnum	Any alphanumeric character
alpha	Any alphabetic character
blank	Any whitespace character that is not a line separator
cntrl	Any control character
d	Any decimal digit
digit	Any decimal digit
graph	Any graphical character
lower	Any lowercase character
print	Any printable character
punct	Any punctuation character
s	Any whitespace character
space	Any whitespace character
upper	Any uppercase character
w	Any word character (alphanumeric characters plus the underscore)
xdigit	Any hexadecimal digit character

In a regular expression, a character class name must be bracketed by [: :]. For example, [:digit:] matches a decimal digit. Furthermore, they must be used within a [ ] pair defining a character class.

Several character classes are supported by shorthand notation:

Character Class Abbreviations		
\d	A decimal digit	[[:digit:]]
\s	A space (space, tab, etc.)	[[:space:]]
\w	A letter (a-z) or digit (0-9) or underscore (_)	[_[:alnum:]]
\D	Not \d	[^[:digit:]]
\S	Not \s	[^[:space:]]
\W	Not \w	[^_[:alnum:]]

In addition, languages supporting regular expressions often provide:

Nonstandard (but Common) Character Class Abbreviations		
\l	A lowercase character	[[:lower:]]
\u	An uppercase character	[[:upper:]]
\L	Not \l	[^[:lower:]]
\U	Not \u	[^[:upper:]]

For full portability, use the character class names rather than these abbreviations.

As an example, consider writing a pattern that describes C++ identifiers: an underscore or a letter followed by a possibly empty sequence of letters, digits, or underscores. To illustrate the subtleties involved, I include a few false attempts:

```
[:alpha:][:alnum:]*                // wrong: characters from the set ":alpha" followed by ...
[[:alpha:]][[:alnum:]]*            // wrong: doesn't accept underscore ('_' is not alpha)
([[:alpha:]]|_)[[:alnum:]]*        // wrong: underscore is not part of alnum either

([[:alpha:]]|_)([[:alnum:]]|_)*        // OK, but clumsy
[[:alpha:]_][[:alnum:]_]*              // OK: include the underscore in the character classes
[_[:alpha:]][_[:alnum:]]*              // also OK
[_[:alpha:]]\w*                        // \w is equivalent to [_[:alnum:]]
```

Finally, here is a function that uses the simplest version of **regex_match()** (§10.4.1) to test whether a string is an identifier:

```
bool is_identifier(const string& s)
{
        regex pat {"[_[:alpha:]]\\w*"};  // underscore or letter
                                         // followed by zero or more underscores, letters, or digits
        return regex_match(s,pat);
}
```

Note the doubling of the backslash to include a backslash in an ordinary string literal. Use raw string literals (§10.4)to alleviate problems with special characters. For example:

```
bool is_identifier(const string& s)
{
        regex pat {R"([_[:alpha:]]\w*)"};
        return regex_match(s,pat);
}
```

Here are some examples of patterns:

```
Ax*              // A, Ax, Axxxx
Ax+              // Ax, Axxx          Not A
\d-?\d           // 1-2, 12           Not 1--2
\w{2}-\d{4,5}    // Ab-1234, XX-54321, 22-5432        Digits are in \w
(\d*:)?(\d+)     // 12:3, 1:23, 123, :123       Not 123:
(bs|BS)          // bs, BS            Not bS
[aeiouy]         // a, o, u           An English vowel, not x
[^aeiouy]        // x, k              Not an English vowel, not e
[a^eiouy]        // a, ^, o, u        An English vowel or ^
```

A **group** (a subpattern) potentially to be represented by a **sub_match** is delimited by parentheses. If you need parentheses that should not define a subpattern, use **(?:** rather than plain **(**. For example:

```
(\s|:|,)*(\d*)      // optional spaces, colons, and/or commas followed by an optional number
```

Assuming that we were not interested in the characters before the number (presumably separators), we could write:

```
(?:\s|:|,)*(\d*)    // optional spaces, colons, and/or commas followed by an optional number
```

This would save the regular expression engine from having to store the first characters: the **(?:** variant has only one subpattern.

Regular Expression Grouping Examples	
\d*\s\w+	No groups (subpatterns)
(\d*)\s(\w+)	Two groups
(\d*)(\s(\w+))+	Two groups (groups do not nest)
(\s*\w*)+	One group; one or more subpatterns; only the last subpattern is saved as a **sub_match**
<(.*?)>(.*?)</\1>	Three groups; the **\1** means "same as group 1"

That last pattern is useful for parsing XML. It finds tag/end-of-tag markers. Note that I used a non-greedy match (a *lazy match*), **.*?**, for the subpattern between the tag and the end tag. Had I used plain **.***, this input would have caused a problem:

```
Always look on the <b>bright</b> side of <b>life</b>.
```

A *greedy match* for the first subpattern would match the first **<** with the last **>**. That would be correct behavior, but unlikely what the programmer wanted.

For a more exhaustive presentation of regular expressions, see [Friedl,1997].

10.4.3 Iterators（迭代器）

We can define a **regex_iterator** for iterating over a sequence of characters finding matches for a pattern. For example, we can use a **sregex_iterator** (a **regex_iterator<string>**) to output all whitespace-separated words in a **string**:

```
void test()
{
    string input = "aa as; asd ++e^asdf asdfg";
    regex pat {R"(\s+(\w+))"};
    for (sregex_iterator p(input.begin(),input.end(),pat); p!=sregex_iterator{}; ++p)
        cout << (*p)[1] << '\n';
}
```

This outputs:

```
as
asd
asdfg
```

We missed the first word, **aa**, because it has no preceding whitespace. If we simplify the pattern to **R"((\w+))"**, we get

```
aa
as
asd
e
asdf
asdfg
```

A **regex_iterator** is a bidirectional iterator, so we cannot directly iterate over an **istream** (which offers only an input iterator). Also, we cannot write through a **regex_iterator**, and the default **regex_iterator** (**regex_iterator{}**) is the only possible end-of-sequence.

10.5 Advice（建议）

[1] Use std::string to own character sequences; §10.2; [CG: SL.str.1].
[2] Prefer string operations to C-style string functions; §10.1.
[3] Use string to declare variables and members rather than as a base class; §10.2.
[4] Return strings by value (rely on move semantics and copy elision); §10.2, §10.2.1; [CG: F.15].
[5] Directly or indirectly, use substr() to read substrings and replace() to write substrings; §10.2.
[6] A string can grow and shrink, as needed; §10.2.
[7] Use at() rather than iterators or [ ] when you want range checking; §10.2, §10.3.
[8] Use iterators and [ ] rather than at() when you want to optimize speed; §10.2, §10.3.
[9] Use a range-for to safely minimize range checking §10.2, §10.3.
[10] string input doesn't overflow; §10.2, §11.3.
[11] Use c_str() or data() to produce a C-style string representation of a string (only) when you have to; §10.2.
[12] Use a stringstream or a generic value extraction function (such as to<X>) for numeric conversion of strings; §11.7.3.
[13] A basic_string can be used to make strings of characters on any type; §10.2.1.
[14] Use the s suffix for string literals meant to be standard-library strings; §10.3 [CG: SL.str.12].
[15] Use string_view as an argument of functions that needs to read character sequences stored in various ways; §10.3 [CG: SL.str.2].
[16] Use string_span<char> as an argument of functions that needs to write character sequences stored in various ways; §10.3. [CG: SL.str.2] [CG: SL.str.11].
[17] Think of a string_view as a kind of pointer with a size attached; it does not own its characters; §10.3.
[18] Use the sv suffix for string literals meant to be standard-library string_views; §10.3.
[19] Use regex for most conventional uses of regular expressions; §10.4.
[20] Prefer raw string literals for expressing all but the simplest patterns; §10.4.
[21] Use regex_match() to match a complete input; §10.4, §10.4.2.
[22] Use regex_search() to search for a pattern in an input stream; §10.4.1.
[23] The regular expression notation can be adjusted to match various standards; §10.4.2.
[24] The default regular expression notation is that of ECMAScript; §10.4.2.
[25] Be restrained; regular expressions can easily become a write-only language; §10.4.2.
[26] Note that \i for a digit i allows you to express a subpattern in terms of a previous subpattern; §10.4.2.
[27] Use ? to make patterns "lazy"; §10.4.2.
[28] Use regex_iterators for iterating over a stream looking for a pattern; §10.4.3.

11

Input and Output

（输入和输出）

What you see is all you get.
– Brian W. Kernighan

- Introduction
- Output
- Input
- I/O State
- I/O of User-Defined Types
- Formatting
  Stream Formatting; printf()-style Formatting
- Streams
  Standard Streams; File Streams; String Streams; Memory Streams; Synchronized Streams
- C-style I/O
- File System
  Paths; Files and Directories
- Advice

11.1 Introduction（引言）

The I/O stream library provides formatted and unformatted buffered I/O of text and numeric values. It is extensible to support user-defined types exactly like built-in types and type safe.

The file system library provides basic facilities for manipulating files and directories.

An **ostream** converts typed objects to a stream of characters (bytes):

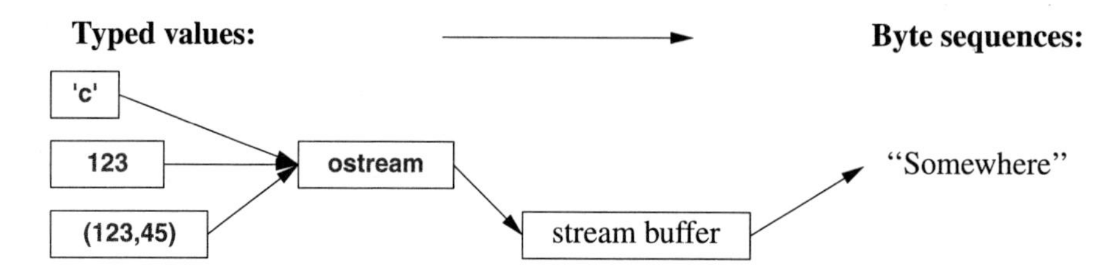

An **istream** converts a stream of characters (bytes) to typed objects:

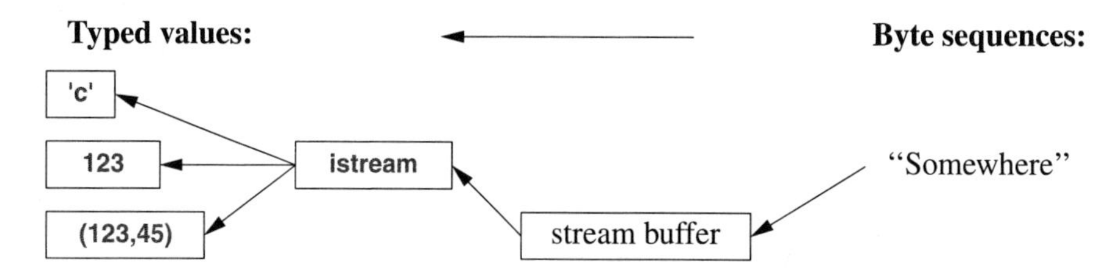

The operations on **istreams** and **ostreams** are described in §11.2 and §11.3. The operations are type-safe, type-sensitive, and extensible to handle user-defined types (§11.5).

Other forms of user interaction, such as graphical I/O, are handled through libraries that are not part of the ISO standard and therefore not described here.

These streams can be used for binary I/O, be used for a variety of character types, be locale specific, and use advanced buffering strategies, but these topics are beyond the scope of this book.

The streams can be used for input into and output from **strings** (§11.3), for formatting into **string** buffers (§11.7.3), into areas of memory (§11.7.4), and for file I/O (§11.9).

The I/O stream classes all have destructors that free all resources owned (such as buffers and file handles). That is, they are examples of "Resource Acquisition Is Initialization" (RAII; §6.3).

11.2 Output（输出）

In **<ostream>**, the I/O stream library defines output for every built-in type. Further, it is easy to define output of a user-defined type (§11.5). The operator << ("put to") is used as an output operator on objects of type **ostream**; **cout** is the standard output stream and **cerr** is the standard stream for reporting errors. By default, values written to **cout** are converted to a sequence of characters. For example, to output the decimal number **10**, we can write:

```
cout << 10;
```

This places the character **1** followed by the character **0** on the standard output stream.

Equivalently, we could write:

```
int x {10};
cout << x;
```

Output of different types can be combined in the obvious way:

```
void h(int i)
{
    cout << "the value of i is ";
    cout << i;
    cout << '\n';
}
```

For h(10), the output will be:

```
the value of i is 10
```

People soon tire of repeating the name of the output stream when outputting several related items. Fortunately, the result of an output expression can itself be used for further output. For example:

```
void h2(int i)
{
    cout << "the value of i is " << i << '\n';
}
```

This h2() produces the same output as h().

A character constant is a character enclosed in single quotes. Note that a character is output as a character rather than as a numerical value. For example:

```
int b = 'b';        // note: char implicitly converted to int
char c = 'c';
cout << 'a' << b << c;
```

The integer value of the character 'b' is 98 (in the ASCII encoding used on the C++ implementation that I used), so this will output a98c.

11.3 Input（输入）

In <istream>, the standard library offers istreams for input. Like ostreams, istreams deal with character string representations of built-in types and can easily be extended to cope with user-defined types.

The operator >> (“get from”) is used as an input operator; cin is the standard input stream. The type of the right-hand operand of >> determines what input is accepted and what is the target of the input operation. For example:

```
int i;
cin >> i;           // read an integer into i

double d;
cin >> d;           // read a double-precision floating-point number into d
```

This reads a number, such as 1234, from the standard input into the integer variable i and a floating-point number, such as 12.34e5, into the double-precision floating-point variable d.

Like output operations, input operations can be chained, so I could equivalently have written:

```
int i;
double d;
cin >> i >> d;          // read into i and d
```

In both cases, the read of the integer is terminated by any character that is not a digit. By default, >> skips initial whitespace, so a suitable complete input sequence would be

```
1234
12.34e5
```

Often, we want to read a sequence of characters. A convenient way of doing that is to read into a string. For example:

```
cout << "Please enter your name\n";
string str;
cin >> str;
cout << "Hello, " << str << "!\n";
```

If you type in Eric the response is:

```
Hello, Eric!
```

By default, a whitespace character, such as a space or a newline, terminates the read, so if you enter Eric Bloodaxe pretending to be the ill-fated king of York, the response is still:

```
Hello, Eric!
```

You can read a whole line using the getline() function. For example:

```
cout << "Please enter your name\n";
string str;
getline(cin,str);
cout << "Hello, " << str << "!\n";
```

With this program, the input Eric Bloodaxe yields the desired output:

```
Hello, Eric Bloodaxe!
```

The newline that terminated the line is discarded, so cin is ready for the next input line.

Using the formatted I/O operations is usually less error-prone, more efficient, and less code than manipulating characters one by one. In particular, istreams take care of memory management and range checking. We can do formatting to and from memory using stringstreams (§11.7.3) or memory streams (§11.7.4).

The standard strings have the nice property of expanding to hold what you put in them; you don't have to pre-calculate a maximum size. So, if you enter a couple of megabytes of semicolons, hello_line() will echo pages of semicolons back at you.

11.4 I/O State（ I/O 状态）

An **iostream** has a state that we can examine to determine whether an operation succeeded. The most common use is to read a sequence of values:

```
vector<int> read_ints(istream& is)
{
	vector<int> res;
	for (int i; is>>i; )
		res.push_back(i);
	return res;
}
```

This reads from **is** until something that is not an integer is encountered. That something will typically be the end of input. What is happening here is that the operation **is>>i** returns a reference to **is**, and testing an **iostream** yields **true** if the stream is ready for another operation.

In general, the I/O state holds all the information needed to read or write, such as formatting information (§11.6.2), error state (e.g., has end-of-input been reached?), and what kind of buffering is used. In particular, a user can set the state to reflect that an error has occurred (§11.5) and clear the state if an error wasn't serious. For example, we could imagine a version of **read_ints()** that accepted a terminating string:

```
vector<int> read_ints(istream& is, const string& terminator)
{
	vector<int> res;
	for (int i; is >> i; )
		res.push_back(i);

	if (is.eof())                    // fine: end of file
		return res;
	if (is.fail()) {                 // we failed to read an int; was it the terminator?
		is.clear();          // reset the state to good()
		string s;
		if (is>>s && s==terminator)
			return res;
		is.setstate(ios_base::failbit);          // add fail() to is's state
	}
	return res;
}

auto v = read_ints(cin,"stop");
```

11.5 I/O of User-Defined Types（用户自定义类型的 I/O）

In addition to the I/O of built-in types and standard **strings**, the **iostream** library allows us to define I/O for our own types. For example, consider a simple type **Entry** that we might use to represent entries in a telephone book:

```
struct Entry {
    string name;
    int number;
};
```

We can define a simple output operator to write an **Entry** using a *{"name",number}* format similar to the one we use for initialization in code:

```
ostream& operator<<(ostream& os, const Entry& e)
{
    return os << "{\"" << e.name << "\", " << e.number << "}";
}
```

A user-defined output operator takes its output stream (by reference) as its first argument and returns it as its result.

The corresponding input operator is more complicated because it has to check for correct formatting and deal with errors:

```
istream& operator>>(istream& is, Entry& e)
    // read { "name" , number } pair. Note: formatted with { " " , and }
{
    char c, c2;
    if (is>>c && c=='{' && is>>c2 && c2=='"') { // start with a { followed by a "
        string name;                    // the default value of a string is the empty string: ""
        while (is.get(c) && c!='"')     // anything before a " is part of the name
            name+=c;

        if (is>>c && c==',') {
            int number = 0;
            if (is>>number>>c && c=='}') { // read the number and a }
                e = {name,number};      // assign to the entry
                return is;
            }
        }
    }

    is.setstate(ios_base::failbit);     // register the failure in the stream
    return is;
}
```

An input operation returns a reference to its **istream** that can be used to test if the operation succeeded. For example, when used as a condition, **is>>c** means "Did we succeed in reading a **char** from **is** into **c**?"

The **is>>c** skips whitespace by default, but **is.get(c)** does not, so this **Entry**-input operator ignores (skips) whitespace outside the name string, but not within it. For example:

```
{ "John Marwood Cleese", 123456        }
{"Michael Edward Palin", 987654}
```

We can read such a pair of values from input into an **Entry** like this:

```
for (Entry ee; cin>>ee; )   // read from cin into ee
    cout << ee << '\n';     // write ee to cout
```

The output is:

```
{"John Marwood Cleese", 123456}
{"Michael Edward Palin", 987654}
```

See §10.4 for a more systematic technique for recognizing patterns in streams of characters (regular expression matching).

11.6 Output Formatting（输出格式化）

The **iostream** and **format** libraries provide operations for controlling the format of input and output. The **iostream** facilities are about as old as C++ and focus on formatting streams of numbers. The **format** facilities (§11.6.2) are recent (C++20) and focus on **printf()**-style (§11.8) specification of the formatting of combinations of values.

Output formatting also offers support for unicode, but that's beyond the scope of this book.

11.6.1 Stream Formatting（流式格式化）

The simplest formatting controls are called *manipulators* and are found in **<ios>**, **<istream>**, **<ostream>**, and **<iomanip>** (for manipulators that take arguments). For example, we can output integers as decimal (the default), octal, or hexadecimal numbers:

```
cout << 1234 << ' ' << hex << 1234 << ' ' << oct << 1234  << dec << 1234 << '\n';  // 1234 4d2 2322 1234
```

We can explicitly set the output format for floating-point numbers:

```
constexpr double d = 123.456;

cout << d << "; "                       // use the default format for d
     << scientific << d << "; "         // use 1.123e2 style format for d
     << hexfloat << d << "; "           // use hexadecimal notation for d
     << fixed << d << "; "              // use 123.456 style format for d
     << defaultfloat << d << '\n';      // use the default format for d
```

This produces:

```
123.456; 1.234560e+002; 0x1.edd2f2p+6; 123.456000; 123.456
```

Precision is an integer that determines the number of digits used to display a floating-point number:

- The *general* format (**defaultfloat**) lets the implementation choose a format that presents a value in the style that best preserves the value in the space available. The precision specifies the maximum number of digits.
- The *scientific* format (**scientific**) presents a value with one digit before a decimal point and an exponent. The precision specifies the maximum number of digits after the decimal point.
- The *fixed* format (**fixed**) presents a value as an integer part followed by a decimal point and a fractional part. The precision specifies the maximum number of digits after the decimal point.

Floating-point values are rounded rather than just truncated, and precision() doesn't affect integer output. For example:

```
cout.precision(8);
cout << "precision(8): " << 1234.56789 << ' ' << 1234.56789 << ' ' << 123456 << '\n';

cout.precision(4);
cout << "precision(4): " << 1234.56789 << ' ' << 1234.56789 << ' ' << 123456 << '\n';
cout << 1234.56789 << '\n';
```

This produces:

```
precision(8): 1234.5679 1234.5679 123456
precision(4): 1235 1235 123456
1235
```

These floating-point manipulators are "sticky"; that is, their effects persist for subsequent floating-point operations. That is, they are designed primarily for formatting streams of values.

We can also specify the size of the field that a number is to be placed into and its alignment in that field.

In addition to the basic numbers, << can also handle time and dates: duration, time_point year_month_date, weekday, month, and zoned_time (§16.2). For example:

```
cout << "birthday: " << November/28/2021 << '\n';
cout << << "zt: " << zoned_time{current_zone(), system_clock::now()} << '\n';
```

This produced:

```
birthday: 2021-11-28
zt: 2021-12-05 11:03:13.5945638 EST
```

The standard also defines << for complex numbers, bitsets (§15.3.2), error codes, and pointers. Stream I/O is extensible, so we can define << for our own (user-defined) types (§11.5).

11.6.2 printf()-style Formatting （printf()风格的格式化）

It has been credibly argued that printf() is the most popular function in C and a significant factor in its success. For example:

```
printf("an int %g and a string '%s'\n",123,"Hello!");
```

This "format string followed by arguments"-style was adopted into C from BCPL and has been followed by many languages. Naturally, printf() has always been part of the C++ standard library, but it suffers from lack of type safety and lack of extensibility to handle user-defined types.

In <format>, the standard library provides a type-safe, though not extensible, printf()-style formatting mechanism. The basic function, format() produces a string:

```
string s = format("Hello, {}\n", val);
```

"Ordinary characters" in the *format string* are simply put into the output string. A *format string* delimited by { and } specify how arguments following the format string are to be inserted into the string. The simplest format string is the empty string, {}, that takes the next argument from the argument list and prints it according to its << default (if any). So, if val is "World", we get the iconic

"Hello, World\n". If val is 127 we get "Hello, 127\n".

The most common use of format() is to output its result:

```
cout << format("Hello, {}\n", val);
```

To see how this works, let's first repeat the examples from (§11.6.1):

```
cout << format("{} {:x} {:o} {:d} {:b}\n", 1234,1234,1234,1234,1234);
```

This gives the same output as the integer example in §11.6.1, except that I added b for binary which is not directly supported by ostream:

```
1234 4d2 2322 1234 10011010010
```

A formatting directive is preceded by a colon. The integer formatting alternatives are x for hexadecimal, o for octal, d for decimal, and b for binary.

By default, format() takes its arguments in order. However, we can specify an arbitrary order. For example:

```
cout << format("{3:} {1:x} {2:o} {0:b}\n", 000, 111, 222, 333);
```

This prints 333 6f 336 0. The number before the colon is the number of the argument to be formatted. In the best C++ style, the numbering starts at zero. This allows us to format an argument more than once:

```
cout << format("{0:} {0:x} {0:o} {0:d} {0:b}\n", 1234); // default, hexadecimal, octal, decimal, binary
```

The ability to place arguments into the output "out of order" is highly praised by people composing messages in different natural languages.

The floating-point formats are the same as for ostream: e for scientific, a for hexfloat, f for fixed, and g for default. For example:

```
cout << format("{0:}; {0:e}; {0:a}; {0:f}; {0:g}\n",123.456);      // default, scientific, hexfloat, fixed, default
```

The result was identical to that from ostream except that the hexadecimal number was not preceded by 0x:

```
123.456; 1.234560e+002; 1.edd2f2p+6; 123.456000; 123.456
```

A dot precedes a precision specifier:

```
cout << format("precision(8): {:.8} {} {}\n", 1234.56789, 1234.56789, 123456);
cout << format("precision(4): {:.4} {} {}\n", 1234.56789, 1234.56789, 123456);
cout << format("{}\n", 1234.56789);
```

Unlike for streams, specifiers are not "sticky" so we get:

```
precision(8): 1234.5679 1234.56789 123456
precision(4): 1235 1234.56789 123456
1234.56789
```

As with stream formatters, we can also specify the size of the field into which a number is to be placed and its alignment in that field.

Like stream formatters, format() can also handle time and dates (§16.2.2). For example:

```
cout << format("birthday: {}\n",November/28/2021);
cout << format("zt: {}", zoned_time{current_zone(), system_clock::now()});
```

As usual, the default formatting of values is identical to that of default stream output formatting. However, format() offers a mini-language of about 60 format specifiers allowing very detailed control over formatting of numbers and dates. For example:

```
auto ymd = 2021y/March/30 ;
cout << format("ymd: {3:%A},{1:} {2:%B},{0:}\n", ymd.year(), ymd.month(), ymd.day(), weekday(ymd));
```

This produced:

```
ymd: Tuesday, March 30, 2021
```

All time and date format strings start with %.

The flexibility offered by the many format specifiers can be important, but it comes with many opportunities for mistakes. Some specifiers come with optional or locale dependent semantics. If a formatting error is caught at run time, a format_error exception is thrown. For example:

```
string ss = format("{:%F}", 2);        // error: mismatched argument; potentially caught at compile time
string sss = format("{%F}", 2);        // error: bad format; potentially caught at compile time
```

The examples so far have had constant formats that can be checked at compile time. The complimentary function vformat() takes a variable as a format to significantly increase flexibility and the opportunities for run-time errors:

```
string fmt = "{}";
cout << vformat(fmt, make_format_args(2));  // OK
fmt = "{:%F}";
cout << vformat(fmt, make_format_args(2));  // error: format and argument mismatch; caught at run time
```

Finally, a formatter can also write directly into a buffer defined by an iterator. For example:

```
string buf;
format_to(back_inserter(buf), "iterator: {} {}\n", "Hi! ", 2022);
cout << buf;    // iterator: Hi! 2022
```

This gets interesting for performance if we use a stream's buffer directly or the buffer for some other output device.

11.7 Streams（流）

The standard library directly supports

- *Standard streams*: streams attached to the system's standard I/O streams (§11.7.1)
- *File streams*: streams attached to files (§11.7.2)
- *String streams*: streams attached to strings (§11.7.3)
- *Memory streams*: stream attached to specific areas of memory (§11.7.4)
- *Synchronized streams*: streams that can be used from multiple threads without data races (§11.7.5)

In addition, we can define our own streams, e.g., attached to communication channels.

Streams cannot be copied; always pass them by reference.

All the standard-library streams are templates with their character type as a parameter. The versions with the names I use here take **chars**. For example, **ostream** is **basic_ostream<char>**. For each such stream, the standard library also provides a version for **wchar_ts**. For example, **wostream** is **basic_ostream<wchar_t>**. The wide character streams can be used for unicode characters.

11.7.1 Standard Streams （标准流）

The standard streams are

- **cout** for "ordinary output"
- **cerr** for unbuffered "error output"
- **clog** for buffered "logging output"
- **cin** for standard input.

11.7.2 File Streams （文件流）

In **<fstream>**, the standard library provides streams to and from a file:

- **ifstreams** for reading from a file
- **ofstreams** for writing to a file
- **fstreams** for reading from and writing to a file

For example:

```
ofstream ofs {"target"};            // "o" for "output"
if (!ofs)
      error("couldn't open 'target' for writing");
```

Testing that a file stream has been properly opened is usually done by checking its state.

```
ifstream ifs {"source"};            // "i" for "input"
if (!ifs)
      error("couldn't open 'source' for reading");
```

Assuming that the tests succeeded, **ofs** can be used as an ordinary **ostream** (just like **cout**) and **ifs** can be used as an ordinary **istream** (just like **cin**).

File positioning and more detailed control of the way a file is opened is possible, but beyond the scope of this book.

For the composition of file names and file system manipulation, see §11.9.

11.7.3 String Streams （字符串流）

In **<sstream>**, the standard library provides streams to and from a **string**:

- **istringstreams** for reading from a **string**
- **ostringstreams** for writing to a **string**
- **stringstreams** for reading from and writing to a **string**.

For example:

```
void test()
{
      ostringstream oss;

      oss << "{temperature," << scientific << 123.4567890 << "}";
      cout << oss.view() << '\n';
}
```

The contents of an **ostringstream** can be read using **str()** (a **string** copy of the contents) or **view()** (a **string_view** of the contents). One common use of an **ostringstream** is to format before giving the resulting string to a GUI. Similarly, a string received from a GUI can be read using formatted input operations (§11.3) by putting it into an **istringstream**.

A **stringstream** can be used for both reading and writing. For example, we can define an operation that can convert any type with a **string** representation into another that can also be represented as a **string**:

```
template<typename Target =string, typename Source =string>
Target to(Source arg)           // convert Source to Target
{
   stringstream buf;
   Target result;

   if (!(buf << arg)                   // write arg into stream
      || !(buf >> result)              // read result from stream
      || !(buf >> std::ws).eof())      // is anything left in stream?
      throw runtime_error{"to<>() failed"};

   return result;
}
```

A function template argument needs to be explicitly mentioned only if it cannot be deduced or if there is no default (§8.2.4), so we can write:

```
auto x1 = to<string,double>(1.2);   // very explicit (and verbose)
auto x2 = to<string>(1.2);          // Source is deduced to double
auto x3 = to<>(1.2);                // Target is defaulted to string; Source is deduced to double
auto x4 = to(1.2);                  // the <> is redundant;
                                    // Target is defaulted to string; Source is deduced to double
```

If all function template arguments are defaulted, the <> can be left out.

I consider this a good example of the generality and ease of use that can be achieved by a combination of language features and standard-library facilities.

11.7.4 Memory Streams（内存流）

From the earliest days of C++, there have been streams attached to sections of memory designated by the user, so that we can read/write directly from/to them. The oldest such streams, **strstream**, have been deprecated for decades, but their replacement, **spanstream**, **ispanstream**, and **ospanstream**, will not become official before C++23. However, they are already widely available;

try your implementation or search GitHub.

An **ospanstream** behaves like an **ostringstream** (§11.7.3) and is initialized like it except that the **ospanstream** takes a **span** rather than a **string** as an argument. For example:

```
void user(int arg)
{
	array<char,128> buf;
	ospanstream ss(buf);
	ss << "write " << arg << " to memory\n";
	// ...
}
```

Attempts overflow the target buffer sets the string state to **failure** (§11.4).

Similarly, an **ispanstream** is like an **istringstream**.

11.7.5 Synchronized Streams （同步流）

In a multi-threaded system, I/O becomes an unreliable mess unless either

- Only one **thread** uses the stream.
- Access to a stream is synchronized so that only one **thread** at a time gains access.

An **osyncstream** guarantees that a sequence of output operations will complete and their results will be as expected in the output buffer even if some other **thread** tries to write. For example:

```
void unsafe(int x, string& s)
{
	cout << x;
	cout << s;
}
```

A different **thread** may introduce a data race (§18.2) and lead to surprising output. An **osyncstream** can be used to avoid that

```
void safer(int x, string& s)
{
	osyncstream oss(cout);
	oss << x;
	oss << s;
}
```

Other **threads** that also use **osyncstreams** will not interfere. Another **thread** that uses **cout** directly could interfere, so either use **ostringstreams** consistently or make sure that only a single **thread** produces output to a specific output stream.

Concurrency can be tricky, so be careful (Chapter 18). Avoid data sharing between **threads** whenever feasible.

11.8 C-style I/O （C 风格的 I/O）

The C++ standard library also supports the C standard-library I/O, including **printf()** and **scanf()**. Many uses of this library are unsafe from a type and security point-of-view, so I don't recommend its use. In particular, it can be difficult to use for safe and convenient input. It does not support user-defined types. If you *don't* use C-style I/O but care about I/O performance, call

```
ios_base::sync_with_stdio(false);        // avoid significant overhead
```

Without that call, the standard **iostreams** (e.g., **cin** and **cout**) can be significantly slowed down to be compatible with the C-style I/O.

If you like **printf()**-style formatted output, use **format** (§11.6.2); it's type safe, easier to use, as flexible, and as fast.

11.9 File System（文件系统）

Most systems have a notion of a *file system* providing access to permanent information stored as *files*. Unfortunately, the properties of file systems and the ways of manipulating them vary greatly. To deal with that, the file system library in **<filesystem>** offers a uniform interface to most facilities of most file systems. Using **<filesystem>**, we can portably

- express file system paths and navigate through a file system
- examine file types and the permissions associated with them

The filesystem library can handle unicode, but explaining how is beyond the scope of this book. I recommend the cppreference [Cppreference] and the Boost filesystem documentation [Boost] for detailed information.

11.9.1 Paths（路径）

Consider an example:

```
path f = "dir/hypothetical.cpp";        // naming a file

assert(exists(f));        // f must exist

if (is_regular_file(f))        // is f an ordinary file?
    cout << f << " is a file; its size is " << file_size(f) << '\n';
```

Note that a program manipulating a file system is usually running on a computer together with other programs. Thus, the contents of a file system can change between two commands. For example, even though we first of all carefully asserted that **f** existed, that may no longer be true when on the next line, we ask if **f** is a regular file.

A **path** is quite a complicated class, capable of handling the varied character sets and conventions of many operating systems. In particular, it can handle file names from command lines as presented by **main()**; for example:

```
int main(int argc, char* argv[])
{
      if (argc < 2) {
            cerr << "arguments expected\n";
            return 1;
      }

      path p {argv[1]};     // create a path from the command line

      cout << p << " " << exists(p) << '\n';       // note: a path can be printed like a string
      // ...
}
```

A path is not checked for validity until it is used. Even then, its validity depends on the conventions of the system on which the program runs.

Naturally, a path can be used to open a file

```
void use(path p)
{
      ofstream f {p};
      if (!f) error("bad file name: ", p);
      f << "Hello, file!";
}
```

In addition to path, <filesystem> offers types for traversing directories and inquiring about the properties of the files found:

File System Types (partial)	
path	A directory path
filesystem_error	A file system exception
directory_entry	A directory entry
directory_iterator	For iterating over a directory
recursive_directory_iterator	For iterating over a directory and its subdirectories

Consider a simple, but not completely unrealistic, example:

```
void print_directory(path p)     // print the names of all files in p
try
{
      if (is_directory(p)) {
            cout << p << ":\n";
            for (const directory_entry& x : directory_iterator{p})
                  cout << "    " << x.path() << '\n';
      }
}
catch (const filesystem_error& ex) {
      cerr << ex.what() << '\n';
}
```

A string can be implicitly converted to a path so we can exercise print_directory like this:

```
void use()
{
      print_directory(".");          // current directory
      print_directory("..");         // parent directory
      print_directory("/");          // Unix root directory
      print_directory("c:");         // Windows volume C

      for (string s; cin>>s; )
            print_directory(s);
}
```

Had I wanted to list subdirectories also, I would have said **recursive_directory_iterator{p}**. Had I wanted to print entries in lexicographical order, I would have copied the **paths** into a **vector** and sorted that before printing.

Class **path** offers many common and useful operations:

Path Operations (partial) **p** and **p2** are **paths**	
value_type	Character type used by the native encoding of the filesystem: **char** on POSIX, **wchar_t** on Windows
string_type	**std::basic_string<value_type>**
const_iterator	A **const** BidirectionalIterator with a **value_type** of **path**
iterator	Alias for **const_iterator**
p=p2	Assign **p2** to **p**
p/=p2	**p** and **p2** concatenated using the file-name separator (by default **/**)
p+=p2	**p** and **p2** concatenated (no separator)
s=p.native()	A reference to the native format of **p**
s=p.string()	**p** in the native format of **p** as a **string**
s=p.generic_string()	**p** in the generic format as a **string**
p2=p.filename()	The filename part of **p**
p2=p.stem()	The stem part of **p**
p2=p.extension()	The file extension part of **p**
i=p.begin()	The beginning iterator of **p**'s element sequence
i= p.end()	The end iterator of **p**'s element sequence
p==p2, p!=p2	Equality and inequality for **p** and **p2**
p<p2, p<=p2, p>p2, p>=p2	Lexicographical comparisons
is>>p, os<<p	Stream I/O to/from **p**
u8path(s)	A path from a UTF-8 encoded source **s**

For example:

```
void test(path p)
{
    if (is_directory(p)) {
        cout << p << ":\n";
        for (const directory_entry& x : directory_iterator(p)) {
            const path& f = x;    // refer to the path part of a directory entry
            if (f.extension() == ".exe")
                cout << f.stem() << " is a Windows executable\n";
            else {
                string n = f.extension().string();
                if (n == ".cpp" || n == ".C" || n == ".cxx")
                    cout << f.stem() << " is a C++ source file\n";
            }
        }
    }
}
```

We use a **path** as a string (e.g., **f.extension**) and we can extract strings of various types from a **path** (e.g., **f.extension().string()**).

Naming conventions, natural languages, and string encodings are rich in complexity. The standard-library filesystem abstractions offer portability and great simplification.

11.9.2 Files and Directories（文件和目录）

Naturally, a file system offers many operations and naturally, different operating systems offer different sets of operations. The standard library offers a few that can be implemented reasonably on a wide variety of systems.

File System Operations (partial) **p**, **p1**, and **p2** are **path**s; **e** is an **error_code**; **b** is a bool indicating success or failure	
exists(p)	Does **p** refer to an existing file system object?
copy(p1,p2)	Copy files or directories from **p1** to **p2**; report errors as exceptions
copy(p1,p2,e)	Copy files or directories; report errors as error codes
b=copy_file(p1,p2)	Copy file contents from **p1** to **p2**; report errors as exceptions
b=create_directory(p)	Create new directory named **p**; all intermediate directories on **p** must exist
b=create_directories(p)	Create new directory named **p**; create all intermediate directories on **p**
p=current_path()	**p** is the current working directory
current_path(p)	Make **p** the current working directory
s=file_size(p)	**s** is the number of bytes in **p**
b=remove(p)	Remove **p** if it is a file or an empty directory

Many operations have overloads that take extra arguments, such as operating-system permissions. The handling of such is far beyond the scope of this book, so look them up if you need them.

Like **copy()**, all operations come in two versions:

- The basic version as listed in the table, e.g., **exists(p)**. The function will throw **filesystem_error** if the operation failed.

- A version with an extra error_code argument, e.g., exists(p,e). Check e to see if the operations succeeded.

We use the error codes when operations are expected to fail frequently in normal use and the throwing operations when an error is considered exceptional.

Often, using an inquiry function is the simplest and most straightforward approach to examining the properties of a file. The <filesystem> library knows about a few common kinds of files and classifies the rest as "other":

File types f is a path or a file_status	
is_block_file(f)	Is f a block device?
is_character_file(f)	Is f a character device?
is_directory(f)	Is f a directory?
is_empty(f)	Is f an empty file or directory?
is_fifo(f)	Is f a named pipe?
is_other(f)	Is f some other kind of file?
is_regular_file(f)	Is f a regular (ordinary) file?
is_socket(f)	Is f a named IPC socket?
is_symlink(f)	Is f a symbolic link?
status_known(f)	Is f's file status known?

11.10 Advice（建议）

[1] iostreams are type-safe, type-sensitive, and extensible; §11.1.
[2] Use character-level input only when you have to; §11.3; [CG: SL.io.1].
[3] When reading, always consider ill-formed input; §11.3; [CG: SL.io.2].
[4] Avoid endl (if you don't know what endl is, you haven't missed anything); [CG: SL.io.50].
[5] Define << and >> for user-defined types with values that have meaningful textual representations; §11.1, §11.2, §11.3.
[6] Use cout for normal output and cerr for errors; §11.1.
[7] There are iostreams for ordinary characters and wide characters, and you can define an iostream for any kind of character; §11.1.
[8] Binary I/O is supported; §11.1.
[9] There are standard iostreams for standard I/O streams, files, and strings; §11.2, §11.3, §11.7.2, §11.7.3.
[10] Chain << operations for a terser notation; §11.2.
[11] Chain >> operations for a terser notation; §11.3.
[12] Input into strings does not overflow; §11.3.
[13] By default >> skips initial whitespace; §11.3.
[14] Use the stream state fail to handle potentially recoverable I/O errors; §11.4.
[15] We can define << and >> operators for our own types; §11.5.
[16] We don't need to modify istream or ostream to add new << and >> operators; §11.5.

[17] Use manipulators or format() to control formatting; §11.6.1, §11.6.2.
[18] precision() specifications apply to all following floating-point output operations; §11.6.1.
[19] Floating-point format specifications (e.g., scientific) apply to all following floating-point output operations; §11.6.1.
[20] #include <ios> or <iostream> when using standard manipulators; §11.6.
[21] Stream formatting manipulators are "sticky" for use for many values in a stream; §11.6.1.
[22] #include <iomanip> when using standard manipulators taking arguments; §11.6.
[23] We can output time, dates, etc. in standard formats; §11.6.1, §11.6.2.
[24] Don't try to copy a stream: streams are move only; §11.7.
[25] Remember to check that a file stream is attached to a file before using it; §11.7.2.
[26] Use stringstreams or memory streams for in-memory formatting; §11.7.3; §11.7.4.
[27] We can define conversions between any two types that both have string representation; §11.7.3.
[28] C-style I/O is not type-safe; §11.8.
[29] Unless you use printf-family functions call ios_base::sync_with_stdio(false); §11.8; [CG: SL.io.10].
[30] Prefer <filesystem> to direct use of platform-specific interfaces; §11.9.

12

Containers

（容器）

It was new. It was singular. It was simple.
It must succeed!
– H. Nelson

12.1 Introduction（引言）

Most computing involves creating collections of values and then manipulating such collections. Reading characters into a **string** and printing out the **string** is a simple example. A class with the main purpose of holding objects is commonly called a *container*. Providing suitable containers for a given task and supporting them with useful fundamental operations are important steps in the construction of any program.

To illustrate the standard-library containers, consider a simple program for keeping names and telephone numbers. This is the kind of program for which different approaches appear "simple and obvious" to people of different backgrounds. The **Entry** class from §11.5 can be used to hold a simple phone book entry. Here, we deliberately ignore many real-world complexities, such as the fact that many phone numbers do not have a simple representation as a 32-bit **int**.

12.2 vector

The most useful standard-library container is **vector**. A **vector** is a sequence of elements of a given type. The elements are stored contiguously in memory. A typical implementation of **vector** (§5.2.2, §6.2) will consist of a handle holding pointers to the first element, one-past-the-last element, and one-past-the-last allocated space (§13.1) (or the equivalent information represented as a pointer plus offsets):

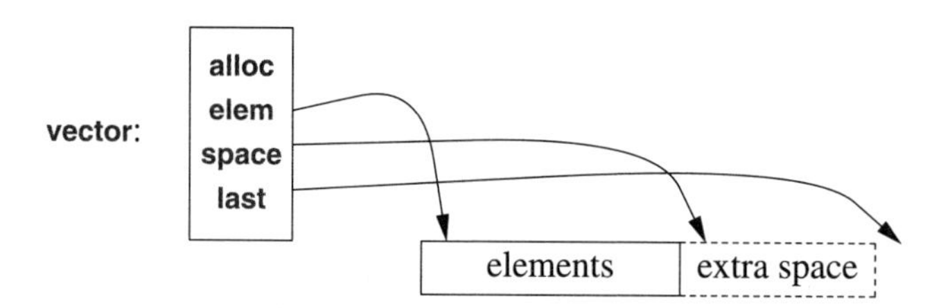

In addition, it holds an allocator (here, **alloc**), from which the **vector** can acquire memory for its elements. The default allocator uses **new** and **delete** to acquire and release memory (§12.7). Using a slightly advanced implementation technique, we can avoid storing any data for simple allocators in a **vector** object.

We can initialize a **vector** with a set of values of its element type:

```
vector<Entry> phone_book = {
      {"David Hume",123456},
      {"Karl Popper",234567},
      {"Bertrand Arthur William Russell",345678}
};
```

Elements can be accessed through subscripting. So, assuming that we have defined **<<** for **Entry**, we can write:

```
void print_book(const vector<Entry>& book)
{
      for (int i = 0; i!=book.size(); ++i)
            cout << book[i] << '\n';
}
```

As usual, indexing starts at **0** so that **book[0]** holds the entry for **David Hume**. The **vector** member function **size()** gives the number of elements.

The elements of a **vector** constitute a range, so we can use a range-**for** loop (§1.7):

```
void print_book(const vector<Entry>& book)
{
      for (const auto& x : book)      // for "auto" see §1.4
            cout << x << '\n';
}
```

When we define a **vector**, we give it an initial size (initial number of elements):

```
vector<int> v1 = {1, 2, 3, 4};           // size is 4
vector<string> v2;                       // size is 0
vector<Shape*> v3(23);                   // size is 23; initial element value: nullptr
vector<double> v4(32,9.9);               // size is 32; initial element value: 9.9
```

An explicit size is enclosed in ordinary parentheses, for example, **(23)**, and by default, the elements are initialized to the element type's default value (e.g., **nullptr** for pointers and **0** for numbers). If you don't want the default value, you can specify one as a second argument (e.g., **9.9** for the **32** elements of **v4**).

The initial size can be changed. One of the most useful operations on a **vector** is **push_back()**, which adds a new element at the end of a **vector**, increasing its size by one. For example, assuming that we have defined **>>** for **Entry**, we can write:

```
void input()
{
      for (Entry e; cin>>e; )
            phone_book.push_back(e);
}
```

This reads **Entry**s from the standard input into **phone_book** until either the end-of-input (e.g., the end of a file) is reached or the input operation encounters a format error.

The standard-library **vector** is implemented so that growing a **vector** by repeated **push_back()**s is efficient. To show how, consider an elaboration of the simple **Vector** from Chapter 5 and Chapter 7 using the representation indicated in the diagram above:

```
template<typename T>
class Vector {
      allocator<T> alloc;  // standard-library allocator of space for Ts
      T* elem;             // pointer to first element
      T* space;            // pointer to first unused (and uninitialized) slot
      T* last;             // pointer to last slot
public:
      // ...
      int size() const { return space-elem; }          // number of elements
      int capacity() const { return last-elem; }       // number of slots available for elements
      // ...
      void reserve(int newsz);              // increase capacity() to newsz
      // ...
      void push_back(const T& t);           // copy t into Vector
      void push_back(T&& t);                // move t into Vector
};
```

The standard-library **vector** has members **capacity()**, **reserve()**, and **push_back()**. The **reserve()** is used by users of **vector** and other **vector** members to make room for more elements. It may have to allocate new memory and when it does, it moves the elements to the new allocation. When **reserve()** moves elements to a new and larger allocation, any pointers to those elements will now point to the wrong location; they have become *invalidated* and should not be used.

Given **capacity()** and **reserve()**, implementing **push_back()** is trivial:

```
template<typename T>
void Vector<T>::push_back(const T& t)
{
     if (capacity()<=size())                  // make sure we have space for t
          reserve(size()==0?8:2*size()); // double the capacity
     construct_at(space,t);                   // initialize *space to t ("place t at space")
     ++space;
}
```

Now allocation and relocation of elements happens only infrequently. I used to use **reserve()** to try to improve performance, but that turned out to be a waste of effort: the heuristic used by **vector** is on average better than my guesses, so now I only explicitly use **reserve()** to avoid reallocation of elements when I want to use pointers to elements.

A **vector** can be copied in assignments and initializations. For example:

```
vector<Entry> book2 = phone_book;
```

Copying and moving **vector**s are implemented by constructors and assignment operators as described in §6.2. Assigning a **vector** involves copying its elements. Thus, after the initialization of **book2**, **book2** and **phone_book** hold separate copies of every **Entry** in the phone book. When a **vector** holds many elements, such innocent-looking assignments and initializations can be expensive. Where copying is undesirable, references or pointers (§1.7) or move operations (§6.2.2) should be used.

The standard-library **vector** is very flexible and efficient. Use it as your default container; that is, use it unless you have a solid reason to use some other container. If you avoid **vector** because of vague concerns about "efficiency," measure. Our intuition is most fallible in matters of the performance of container uses.

12.2.1 Elements（元素）

Like all standard-library containers, **vector** is a container of elements of some type **T**, that is, a **vector<T>**. Just about any type qualifies as an element type: built-in numeric types (such as **char**, **int**, and **double**), user-defined types (such as **string**, **Entry**, **list<int>**, and **Matrix<double,2>**), and pointers (such as **const char***, **Shape***, and **double***). When you insert a new element, its value is copied into the container. For example, when you put an integer with the value **7** into a container, the resulting element really has the value **7**. The element is not a reference or a pointer to some object containing **7**. This makes for nice, compact containers with fast access. For people who care about memory sizes and run-time performance this is critical.

If you have a class hierarchy (§5.5) that relies on **virtual** functions to get polymorphic behavior, do not store objects directly in a container. Instead store a pointer (or a smart pointer; §15.2.1). For example:

```
vector<Shape> vs;                    // No, don't - there is no room for a Circle or a Smiley (§5.5)
vector<Shape*> vps;                  // better, but see §5.5.3 (don't leak)
vector<unique_ptr<Shape>> vups;      // OK
```

12.2.2 Range Checking（范围检查）

The standard-library **vector** does not guarantee range checking. For example:

```
void silly(vector<Entry>& book)
{
	int i = book[book.size()].number;	// book.size() is out of range
	// ...
}
```

That initialization is likely to place some random value in **i** rather than giving an error. This is undesirable, and out-of-range errors are a common problem. Consequently, I often use a simple range-checking adaptation of **vector**:

```
template<typename T>
struct Vec : std::vector<T> {
	using vector<T>::vector;	// use the constructors from vector (under the name Vec)

	T& operator[](int i) { return vector<T>::at(i); }	// range check
	const T& operator[](int i) const { return vector<T>::at(i); }	// range check const objects; §5.2.1

	auto begin() { return Checked_iter<vector<T>>{*this}; }	// see §13.1
	auto end() { return Checked_iter<vector<T>>{*this, vector<T>::end()}; }
};
```

Vec inherits everything from **vector** except for the subscript operations that it redefines to do range checking. The **at()** operation is a **vector** subscript operation that throws an exception of type **out_of_range** if its argument is out of the **vector**'s range (§4.2).

For **Vec**, an out-of-range access will throw an exception that the user can catch. For example:

```
void checked(Vec<Entry>& book)
{
	try {
		book[book.size()] = {"Joe",999999};	// will throw an exception
		// ...
	}
	catch (out_of_range&) {
		cerr << "range error\n";
	}
}
```

The exception will be thrown, and then caught (§4.2). If the user doesn't catch an exception, the program will terminate in a well-defined manner rather than proceeding or failing in an undefined manner. One way to minimize surprises from uncaught exceptions is to use a **main()** with a **try**-block as its body. For example:

```
int main()
try {
	// your code
}
```

```
catch (out_of_range&) {
      cerr << "range error\n";
}
catch (...) {
      cerr << "unknown exception thrown\n";
}
```

This provides default exception handlers so that if we fail to catch some exception, an error message is printed on the standard error-diagnostic output stream **cerr** (§11.2).

Why doesn't the standard guarantee range checking? Many performance-critical applications use **vectors** and checking all subscripting implies a cost on the order of 10%. Obviously, that cost can vary dramatically depending on hardware, optimizers, and an application's use of subscripting. However, experience shows that such overhead can lead people to prefer the far more unsafe built-in arrays. Even the mere fear of such overhead can lead to disuse. At least **vector** is easily range checked at debug time and we can build checked versions on top of the unchecked default.

A range-**for** avoids range errors at no cost by implicitly accessing all elements in the range. As long as their arguments are valid, the standard-library algorithms do the same to ensure the absence of range errors.

If you use **vector::at()** directly in your code, you don't need my **Vec** workaround. Furthermore, some standard libraries have range-checked **vector** implementations that offer more complete checking than **Vec**.

12.3 list

The standard library offers a doubly-linked list called **list**:

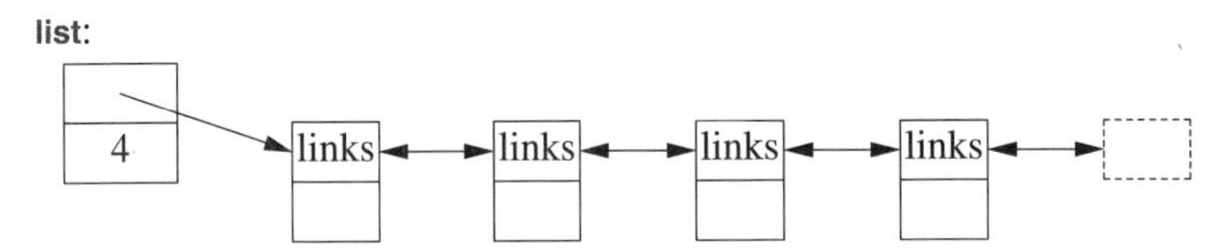

We use a **list** for sequences where we want to insert and delete elements without moving other elements. Insertion and deletion of phone book entries could be common, so a **list** could be appropriate for representing a simple phone book. For example:

```
list<Entry> phone_book = {
      {"David Hume",123456},
      {"Karl Popper",234567},
      {"Bertrand Arthur William Russell",345678}
};
```

When we use a linked list, we tend not to access elements using subscripting the way we commonly do for vectors. Instead, we might search the list looking for an element with a given value. To do this, we take advantage of the fact that a **list** is a sequence as described in Chapter 13:

```
int get_number(const string& s)
{
	for (const auto& x : phone_book)
		if (x.name==s)
			return x.number;
	return 0; // use 0 to represent "number not found"
}
```

The search for s starts at the beginning of the list and proceeds until s is found or the end of phone_book is reached.

Sometimes, we need to identify an element in a list. For example, we may want to delete an element or insert a new element before it. To do that we use an *iterator*: a list iterator identifies an element of a list and can be used to iterate through a list (hence its name). Every standard-library container provides the functions begin() and end(), which return an iterator to the first and to one-past-the-last element, respectively (§13.1). Using iterators explicitly, we can – less elegantly – write the get_number() function like this:

```
int get_number(const string& s)
{
	for (auto p = phone_book.begin(); p!=phone_book.end(); ++p)
		if (p->name==s)
			return p->number;
	return 0; // use 0 to represent "number not found"
}
```

In fact, this is roughly the way the terser and less error-prone range-for loop is implemented by the compiler. Given an iterator p, *p is the element to which it refers, ++p advances p to refer to the next element, and when p refers to a class with a member m, then p->m is equivalent to (*p).m.

Adding elements to a list and removing elements from a list is easy:

```
void f(const Entry& ee, list<Entry>::iterator p, list<Entry>::iterator q)
{
	phone_book.insert(p,ee);	// add ee before the element referred to by p
	phone_book.erase(q);		// remove the element referred to by q
}
```

For a list, insert(p,elem) inserts an element with a copy of the value elem before the element pointed to by p. Here, p may be an iterator pointing one-beyond-the-end of the list. Conversely, erase(p) removes the element pointed to by p and destroys it.

These list examples could be written identically using vector and (surprisingly, unless you understand machine architecture) often perform better with a vector than with a list. When all we want is a sequence of elements, we have a choice between using a vector and a list. Unless you have a reason not to, use a vector. A vector performs better for traversal (e.g., find() and count()) and for sorting and searching (e.g., sort() and equal_range(); §13.5, §15.3.3).

12.4 forward_list

The standard library also offers a singly-linked list called **forward_list**:

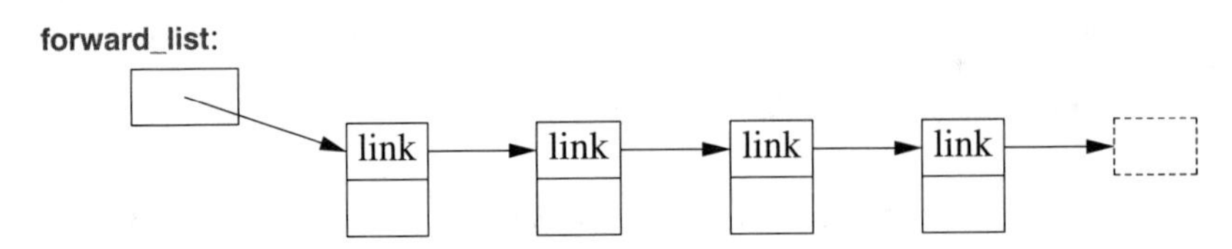

A **forward_list** differs from a (doubly-linked) **list** by only allowing forward iteration. The point of that is to save space. There is no need to keep a predecessor pointer in each link and the size of an empty **forward_list** is just one pointer. A **forward_list** doesn't even keep its number of elements. If you need the number of elements, count. If you can't afford to count, you probably shouldn't use a **forward_list**.

12.5 map

Writing code to look up a name in a list of *(name,number)* pairs is quite tedious. In addition, a linear search is inefficient for all but the shortest lists. The standard library offers a balanced binary search tree (usually a red-black tree) called **map**:

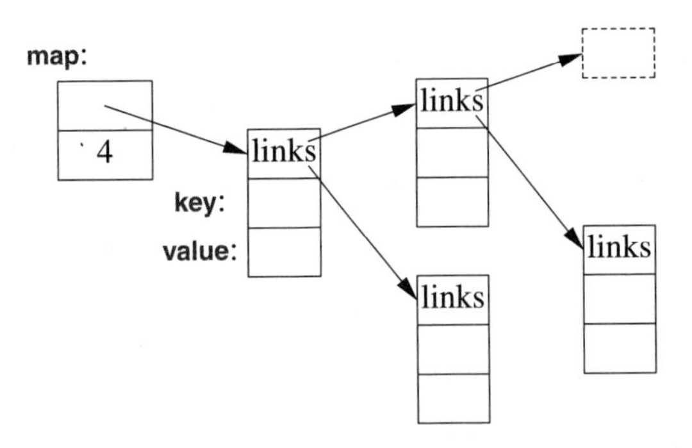

In other contexts, a **map** is known as an associative array or a dictionary.

The standard-library **map** is a container of pairs of values optimized for lookup and insertion. We can use the same initializer as for **vector** and **list** (§12.2, §12.3):

```
map<string,int> phone_book {
      {"David Hume",123456},
      {"Karl Popper",234567},
      {"Bertrand Arthur William Russell",345678}
};
```

When indexed by a value of its first type (called the *key*), a **map** returns the corresponding value of the second type (called the *value* or the *mapped type*). For example:

```
int get_number(const string& s)
{
      return phone_book[s];
}
```

In other words, subscripting a **map** is essentially the lookup we called **get_number()**. If a **key** isn't found, it is entered into the **map** with a default value for its **value**. The default value for an integer type is **0** and that just happens to be a reasonable value to represent an invalid telephone number.

If we wanted to avoid entering invalid numbers into our phone book, we could use **find()** and **insert()** (§12.8) instead of **[ ]**.

12.6 unordered_map

The cost of a **map** lookup is **O(log(n))** where **n** is the number of elements in the **map**. That's pretty good. For example, for a **map** with 1,000,000 elements, we perform only about 20 comparisons and indirections to find an element. However, in many cases, we can do better by using a hashed lookup rather than a comparison using an ordering function, such as **<**. The standard-library hashed containers are referred to as "unordered" because they don't require an ordering function:

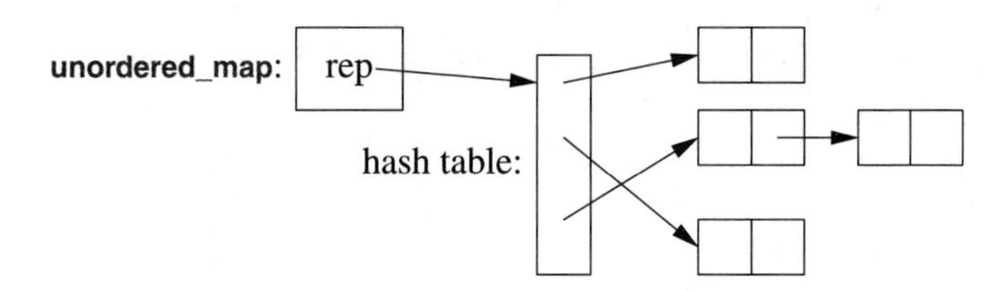

For example, we can use an **unordered_map** from **<unordered_map>** for our phone book:

```
unordered_map<string,int> phone_book {
      {"David Hume",123456},
      {"Karl Popper",234567},
      {"Bertrand Arthur William Russell",345678}
};
```

Like for a **map**, we can subscript an **unordered_map**:

```
int get_number(const string& s)
{
      return phone_book[s];
}
```

The standard library provides a default hash function for **string**s as well as for other built-in and standard-library types. If necessary, we can provide our own. Possibly, the most common need for a custom hash function comes when we want an unordered container of one of our own types. A hash function is often implemented as a function object (§7.3.2). For example:

```
struct Record {
      string name;
      int product_code;
      // ...
};

struct Rhash {         // a hash function for Record
      size_t operator()(const Record& r) const
      {
            return hash<string>()(r.name) ^ hash<int>()(r.product_code);
      }
};

unordered_set<Record,Rhash> my_set; // set of Records using Rhash for lookup
```

Designing good hash functions is an art and often requires knowledge of the data to which it will be applied. Creating a new hash function by combining existing hash functions using exclusive-or (^) is simple and often very effective. However, be careful to ensure that every value that takes part in the hash really helps to distinguish the values. For example, unless you can have several names for the same product code (or several product codes for the same name), combining the two hashes provides no benefits.

We can avoid explicitly passing the **hash** operation by defining it as a specialization of the standard-library **hash**:

```
namespace std {            // make a hash function for Record

      template<> struct hash<Record> {
            using argument_type = Record;
            using result_type = size_t;

            result_type operator()(const Record& r) const
            {
                  return hash<string>()(r.name) ^ hash<int>()(r.product_code);
            }
      };
}
```

Note the differences between a **map** and an **unordered_map**:

- A **map** requires an ordering function (the default is <) and yields an ordered sequence.
- A **unordered_map** requires an equality function (the default is ==); it does not maintain an order among its elements.

Given a good hash function, an **unordered_map** is much faster than a **map** for large containers. However, the worst-case behavior of an **unordered_map** with a poor hash function is far worse than that of a **map**.

12.7 Allocators（分配器）

By default, standard-library containers allocate space using **new**. Operators **new** and **delete** provide a general free store (also called dynamic memory or heap) that can hold objects of arbitrary size and user-controlled lifetime. This implies time and space overheads that can be eliminated in many special cases. Therefore, the standard-library containers offer the opportunity to install allocators with specific semantics where needed. This has been used to address a wide variety of concerns related to performance (e.g., pool allocators), security (allocators that clean-up memory as part of deletion), per-thread allocation, and non-uniform memory architectures (allocating in specific memories with pointer types to match). This is not the place to discuss these important, but very specialized and often advanced techniques. However, I will give one example motivated by a real-world problem for which a pool allocator was the solution.

An important, long-running system used an event queue (see §18.4) using **vector**s as events that were passed as **shared_ptr**s. That way, the last user of an event implicitly deleted it:

```
struct Event {
    vector<int> data = vector<int>(512);
};

list<shared_ptr<Event>> q;

void producer()
{
    for (int n = 0; n!=LOTS; ++n) {
        lock_guard lk {m};          // m is a mutex; see §18.3
        q.push_back(make_shared<Event>());
        cv.notify_one();            // cv is a condition_variable; see §18.4
    }
}
```

From a logical point of view this worked nicely. It is logically simple, so the code is robust and maintainable. Unfortunately, this led to massive fragmentation. After 100,000 events had been passed among 16 producers and 4 consumers, more than 6GB of memory had been consumed.

The traditional solution to fragmentation problems is to rewrite the code to use a pool allocator. A pool allocator is an allocator that manages objects of a single fixed size and allocates space for many objects at a time, rather than using individual allocations. Fortunately, C++ offers direct support for that. The pool allocator is defined in the **pmr** ("polymorphic memory resource") sub-namespace of **std**:

```
pmr::synchronized_pool_resource pool;           // make a pool

struct Event {
    vector<int> data = vector<int>{512,&pool};  // let Events use the pool
};

list<shared_ptr<Event>> q {&pool};              // let q use the pool
```

```
void producer()
{
    for (int n = 0; n!=LOTS; ++n) {
        scoped_lock lk {m};        // m is a mutex (§18.3)
        q.push_back(allocate_shared<Event,pmr::polymorphic_allocator<Event>>{&pool});
        cv.notify_one();
    }
}
```

Now, after 100,000 events had been passed among 16 producers and 4 consumers, less than 3MB of memory had been consumed. That's about a 2000-fold improvement! Naturally, the amount of memory actually in use (as opposed to memory wasted to fragmentation) is unchanged. After eliminating fragmentation, memory use was stable over time so the system could run for months.

Techniques like this have been applied with good effects from the earliest days of C++, but generally they required code to be rewritten to use specialized containers. Now, the standard containers optionally take allocator arguments. The default is for the containers to use **new** and **delete**. Other polymorphic memory resources include

- **unsynchronized_polymorphic_resource**; like **polymorphic_resource** but can only be used by one thread.
- **monotonic_polymorphic_resource**; a fast allocator that releases its memory only upon its destruction and can only be used by one thread.

A polymorphic resource must be derived from **memory_resource** and define members **allocate()**, **deallocate()**, and **is_equal()**. The idea is for users to build their own resources to tune code.

12.8 Container Overview（容器概述）

The standard library provides some of the most general and useful container types to allow the programmer to select a container that best serves the needs of an application:

Standard Container Summary	
vector<T>	A variable-size vector (§12.2)
list<T>	A doubly-linked list (§12.3)
forward_list<T>	A singly-linked list
deque<T>	A double-ended queue
map<K,V>	An associative array (§12.5)
multimap<K,V>	A map in which a key can occur many times
unordered_map<K,V>	A map using a hashed lookup (§12.6)
unordered_multimap<K,V>	A multimap using a hashed lookup
set<T>	A set (a map with just a key and no value)
multiset<T>	A set in which a value can occur many times
unordered_set<T>	A set using a hashed lookup
unordered_multiset<T>	A multiset using a hashed lookup

The unordered containers are optimized for lookup with a key (often a string); in other words, they are hash tables.

The containers are defined in namespace **std** and presented in headers **<vector>**, **<list>**, **<map>**, etc. (§9.3.4). In addition, the standard library provides container adaptors **queue<T>**, **stack<T>**, and **priority_queue<T>**. Look them up if you need them. The standard library also provides more specialized container-like types, such as **array<T,N>** (§15.3.1) and **bitset<N>** (§15.3.2).

The standard containers and their basic operations are designed to be similar from a notational point of view. Furthermore, the meanings of the operations are equivalent for the various containers. Basic operations apply to every kind of container for which they make sense and can be efficiently implemented:

Standard Container Operations (partial)	
value_type	The type of an element
p=c.begin()	p points to first element of c; also cbegin() for an iterator to const
p=c.end()	p points to one-past-the-last element of c; also cend() for an iterator to const
k=c.size()	k is the number of elements in c
c.empty()	Is c empty?
k=c.capacity()	k is the number of elements that c can hold without a new allocation
c.reserve(k)	Increase the capacity to k; if k<=c.capacity(), c.reserve(k) does nothing
c.resize(k)	Make the number of elements k; added elements have the default value value_type{}
c[k]	The kth element of c; zero-based; no range guaranteed checking
c.at(k)	The kth element of c; if out of range, throw out_of_range
c.push_back(x)	Add x at the end of c; increase the size of c by one
c.emplace_back(a)	Add value_type{a} at the end of c; increase the size of c by one
q=c.insert(p,x)	Add x before p in c
q=c.erase(p)	Remove element at p from c
c=c2	Assignment: copy all elements from c2 to get c==c2
b=(c==c2)	Equality of all elements of c and c2; b==true if equal
x=(c<=>c2)	Lexicographical order of c and c2: x<0 if c is less than c2, x==0 if equal, and 0<x if greater than. !=, <, <=, >, and >= are generated from <=>

This notational and semantic uniformity enables programmers to provide new container types that can be used in a very similar manner to the standard ones. The range-checked vector, **Vector** (§4.3, Chapter 5), is an example of that. The uniformity of container interfaces allows us to specify algorithms independently of individual container types. However, each has strengths and weaknesses. For example, subscripting and traversing a **vector** is cheap and easy. On the other hand, **vector** elements are moved to different locations when we insert or remove elements; **list** has exactly the opposite properties. Please note that a **vector** is usually more efficient than a **list** for short sequences of small elements (even for **insert()** and **erase()**). I recommend the standard-library **vector** as the default type for sequences of elements: you need a reason to choose another.

Consider the singly-linked list, **forward_list**, a container optimized for the empty sequence (§12.3). An empty **forward_list** occupies just one word, whereas an empty **vector** occupies three. Empty sequences, and sequences with only an element or two, are surprisingly common and useful.

An emplace operation, such as emplace_back() takes arguments for an element's constructor and builds the object in a newly allocated space in the container, rather than copying an object into the container. For example, for a vector<pair<int,string>> we could write:

```
v.push_back(pair{1,"copy or move"});    // make a pair and move it into v
v.emplace_back(1,"build in place");     // build a pair in v
```

For simple examples like this, optimizations can result in equivalent performance for both calls.

12.9 Advice（建议）

[1] An STL container defines a sequence; §12.2.
[2] STL containers are resource handles; §12.2, §12.3, §12.5, §12.6.
[3] Use vector as your default container; §12.2, §12.8; [CG: SL.con.2].
[4] For simple traversals of a container, use a range-for loop or a begin/end pair of iterators; §12.2, §12.3.
[5] Use reserve() to avoid invalidating pointers and iterators to elements; §12.2.
[6] Don't assume performance benefits from reserve() without measurement; §12.2.
[7] Use push_back() or resize() on a container rather than realloc() on an array; §12.2.
[8] Don't use iterators into a resized vector; §12.2 [CG: ES.65].
[9] Do not assume that [ ] range checks; §12.2.
[10] Use at() when you need guaranteed range checks; §12.2; [CG: SL.con.3].
[11] Use range-for and standard-library algorithms for cost-free avoidance of range errors; §12.2.2.
[12] Elements are copied into a container; §12.2.1.
[13] To preserve polymorphic behavior of elements, store pointers (built-in or user-defined); §12.2.1.
[14] Insertion operations, such as insert() and push_back(), are often surprisingly efficient on a vector; §12.3.
[15] Use forward_list for sequences that are usually empty; §12.8.
[16] When it comes to performance, don't trust your intuition: measure; §12.2.
[17] A map is usually implemented as a red-black tree; §12.5.
[18] An unordered_map is a hash table; §12.6.
[19] Pass a container by reference and return a container by value; §12.2.
[20] For a container, use the ()-initializer syntax for sizes and the {}-initializer syntax for sequences of elements; §5.2.3, §12.2.
[21] Prefer compact and contiguous data structures; §12.3.
[22] A list is relatively expensive to traverse; §12.3.
[23] Use unordered containers if you need fast lookup for large amounts of data; §12.6.
[24] Use ordered containers (e.g., map and set) if you need to iterate over their elements in order; §12.5.
[25] Use unordered containers (e.g., unordered_map) for element types with no natural order (i.e., no reasonable <); §12.5.
[26] Use associative containers (e.g., map and list) when you need pointers to elements to be stable as the size of the container changes; §12.8.

[27] Experiment to check that you have an acceptable hash function; §12.6.
[28] A hash function obtained by combining standard hash functions for elements using the exclusive-or operator (^) is often good; §12.6.
[29] Know your standard-library containers and prefer them to handcrafted data structures; §12.8.
[30] If your application is suffering performance problems related to memory, minimize free store use and/or consider using a specialized allocator; §12.7.

13

Algorithms
（算法）

Do not multiply entities beyond necessity.
– William Occam

13.1 Introduction（引言）

A data structure, such as a list or a vector, is not very useful on its own. To use one, we need operations for basic access such as adding and removing elements (as is provided for list and vector). Furthermore, we rarely just store objects in a container. We sort them, print them, extract subsets, remove elements, search for objects, etc. Consequently, the standard library provides the most common algorithms for containers in addition to providing the most common container types. For example, we can simply and efficiently sort a vector of Entrys and place a copy of each unique vector element on a list:

```
void f(vector<Entry>& vec, list<Entry>& lst)
{
     sort(vec.begin(),vec.end());                          // use < for order
     unique_copy(vec.begin(),vec.end(),lst.begin());       // don't copy adjacent equal elements
}
```

For this to work, less than (<) and equal (==) must be defined for Entrys. For example:

```
bool operator<(const Entry& x, const Entry& y)     // less than
{
      return x.name<y.name;           // order Entries by their names
}
```

A standard algorithm is expressed in terms of (half-open) sequences of elements. A *sequence* is represented by a pair of iterators specifying the first element and the one-beyond-the-last element:

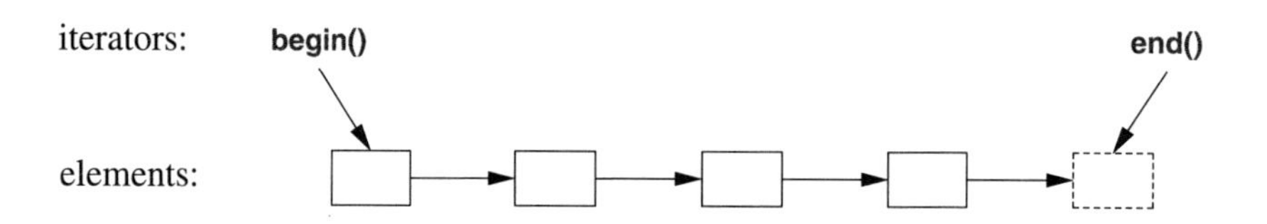

In the example, **sort()** sorts the sequence defined by the pair of iterators **vec.begin()** and **vec.end()**, which just happens to be all the elements of a **vector**. For writing (output), we need only to specify the first element to be written. If more than one element is written, the elements following that initial element will be overwritten. Thus, to avoid errors, **lst** must have at least as many elements as there are unique values in **vec**.

Unfortunately, the standard library doesn't offer an abstraction to support range-checked writing into a container. However, we can define one:

```
template<typename C>
class Checked_iter {
public:
      using value_type = typename C::value_type;
      using difference_type = int;

      Checked_iter() { throw Missing_container{}; } // concept forward_iterator requires a default constructor
      Checked_iter(C& cc) : pc{ &cc } {}
      Checked_iter(C& cc, typename C::iterator pp) : pc{ &cc }, p{ pp } {}

      Checked_iter& operator++() { check_end(); ++p; return *this; }
      Checked_iter operator++(int) { check_end(); auto t{ *this }; ++p; return t; }
      value_type& operator*() const { check_end();  return *p; }

      bool operator==(const Checked_iter& a) const { return p==a.p; }
      bool operator!=(const Checked_iter& a) const { return p!=a.p; }
private:
      void check_end() const { if (p == pc->end()) throw Overflow{}; }
      C* pc {};  // default initialize to nullptr
      typename C::iterator p = pc->begin();
};
```

Obviously, this is not standard-library quality, but it shows the idea:

```
vector<int> v1 {1, 2, 3};          // three elements
vector<int> v2(2);                 // two elements

copy(v1,v2.begin());               // will overflow
copy(v1,Checked_iter{v2});         // will throw
```

If, in the read-and-sort example, we had wanted to place the unique elements in a new list, we could have written:

```
list<Entry> f(vector<Entry>& vec)
{
     list<Entry> res;
     sort(vec.begin(),vec.end());
     unique_copy(vec.begin(),vec.end(),back_inserter(res));       // append to res
     return res;
}
```

The call back_inserter(res) constructs an iterator for res that adds elements at the end of a container, extending the container to make room for them. This saves us from first having to allocate a fixed amount of space and then filling it. Thus, the standard containers plus back_inserter()s eliminate the need to use error-prone, explicit C-style memory management using realloc(). The standard-library list has a move constructor (§6.2.2) that makes returning res by value efficient (even for lists of thousands of elements).

When we find the pair-of-iterators style of code, such as sort(vec.begin(),vec.end()), tedious, we can use range versions of the algorithms and write sort(vec) (§13.5). The two versions are equivalent. Similarly, a range-for loop is roughly equivalent to a C-style loop using iterators directly:

```
for (auto& x : v) cout<<x;                          // write out all elements of v
for (auto p = v.begin(); p!=v.end(); ++p) cout<<*p;    // write out all elements of v
```

In addition to being simpler and less error-prone, the range-for version is often also more efficient.

13.2 Use of Iterators（使用迭代器）

For a container, a few iterators referring to useful elements can be obtained; begin() and end() are the best examples of this. In addition, many algorithms return iterators. For example, the standard algorithm find looks for a value in a sequence and returns an iterator to the element found:

```
bool has_c(const string& s, char c)      // does s contain the character c?
{
     auto p = find(s.begin(),s.end(),c);
     if (p!=s.end())
          return true;
     else
          return false;
}
```

Like many standard-library search algorithms, find returns end() to indicate "not found." An equivalent, shorter, definition of has_c() is:

```
bool has_c(const string& s, char c)        // does s contain the character c?
{
      return find(s,c)!=s.end();
}
```

A more interesting exercise would be to find the location of all occurrences of a character in a string. We can return the set of occurrences as a **vector<char*>**s. Returning a **vector** is efficient because **vector** provides move semantics (§6.2.1). Assuming that we would like to modify the locations found, we pass a non-**const** string:

```
vector<string::iterator> find_all(string& s, char c)        // find all occurrences of c in s
{
      vector<char*> res;
      for (auto p = s.begin(); p!=s.end(); ++p)
            if (*p==c)
                  res.push_back(&*p);
      return res;
}
```

We iterate through the string using a conventional loop, moving the iterator **p** forward one element at a time using **++** and looking at the elements using the dereference operator *. We could test **find_all()** like this:

```
void test()
{
      string m {"Mary had a little lamb"};
      for (auto p : find_all(m,'a'))
            if (*p!='a')
                  cerr << "a bug!\n";
}
```

That call of **find_all()** could be graphically represented like this:

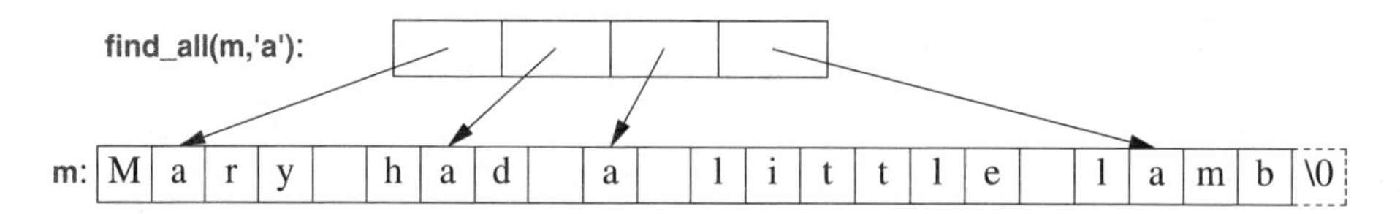

Iterators and standard algorithms work equivalently on every standard container for which their use makes sense. Consequently, we could generalize **find_all()**:

```
template<typename C, typename V>
vector<typename C::iterator> find_all(C& c, V v)        // find all occurrences of v in c
{
      vector<typename C::iterator> res;
      for (auto p = c.begin(); p!=c.end(); ++p)
            if (*p==v)
                  res.push_back(p);
      return res;
}
```

The **typename** is needed to inform the compiler that **C**'s **iterator** is supposed to be a type and not a value of some type, say, the integer **7**.

Alternatively, we could have returned a vector of ordinary pointers to the elements:

```
template<typename C, typename V>
auto find_all(C& c, V v)          // find all occurrences of v in c
{
     vector<range_value_t<C>*> res;
     for (auto& x : c)
          if (x==v)
               res.push_back(&x);
     return res;
}
```

While I was at it, I also simplified the code by using a range-**for** loop and the standard-library **range_value_t** (§16.4.4) to name the type of the elements. A simplified version of **range_value_t** can be defined like this:

```
template<typename T>
using range_value_type_t = T::value_type;
```

Using either version of **find_all()**, we can write:

```
void test()
{
     string m {"Mary had a little lamb"};

     for (auto p : find_all(m,'a'))               // p is a string::iterator
          if (*p!='a')
               cerr << "string bug!\n";

     list<int> ld {1, 2, 3, 1, -11, 2};
     for (auto p : find_all(ld,1))                // p is a list<int>::iterator
          if (*p!=1)
               cerr << "list bug!\n";

     vector<string> vs {"red", "blue", "green", "green", "orange", "green"};
     for (auto p : find_all(vs,"red"))            // p is a vector<string>::iterator
          if (*p!="red")
               cerr << "vector bug!\n";

     for (auto p : find_all(vs,"green"))
          *p = "vert";
}
```

Iterators are used to separate algorithms and containers. An algorithm operates on its data through iterators and knows nothing about the container in which the elements are stored. Conversely, a container knows nothing about the algorithms operating on its elements; all it does is to supply iterators upon request (e.g., **begin()** and **end()**). This model of separation between data storage and algorithm delivers very general and flexible software.

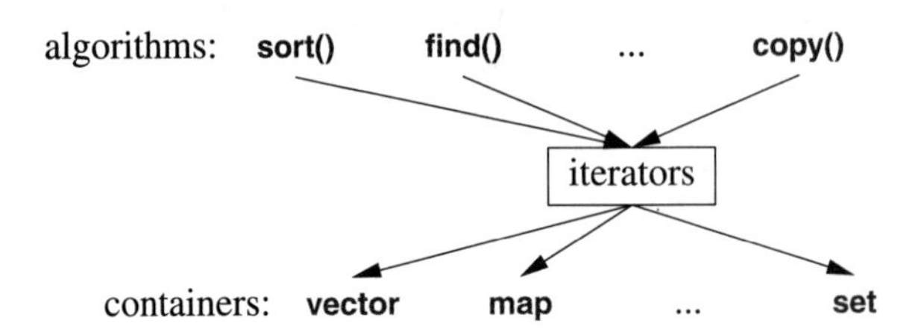

13.3 Iterator Types（迭代器类型）

What are iterators really? Any particular iterator is an object of some type. There are, however, many different iterator types – an iterator needs to hold the information necessary for doing its job for a particular container type. These iterator types can be as different as the containers and the specialized needs they serve. For example, a **vector**'s iterator could be an ordinary pointer, because a pointer is quite a reasonable way of referring to an element of a **vector**:

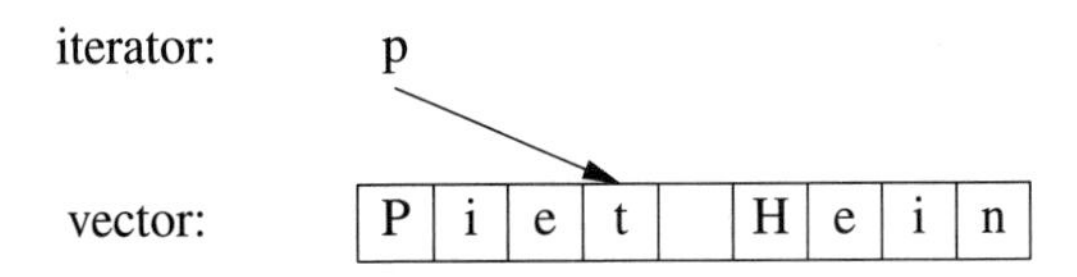

Alternatively, a **vector** iterator could be implemented as a pointer to the **vector** plus an index:

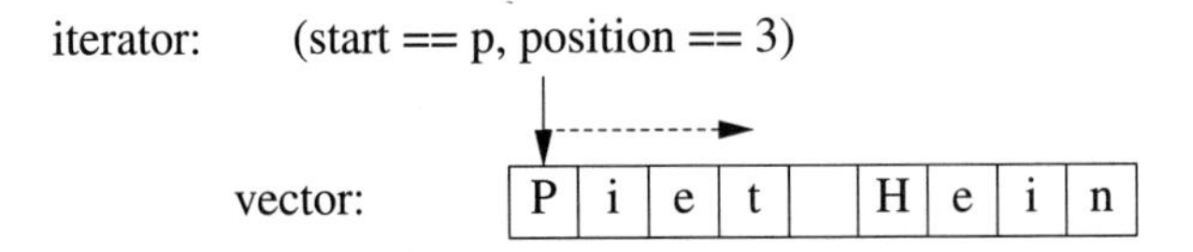

Using such an iterator allows range checking.

A **list** iterator must be something more complicated than a simple pointer to an element because an element of a **list** in general does not know where the next element of that **list** is. Thus, a **list** iterator might be a pointer to a link:

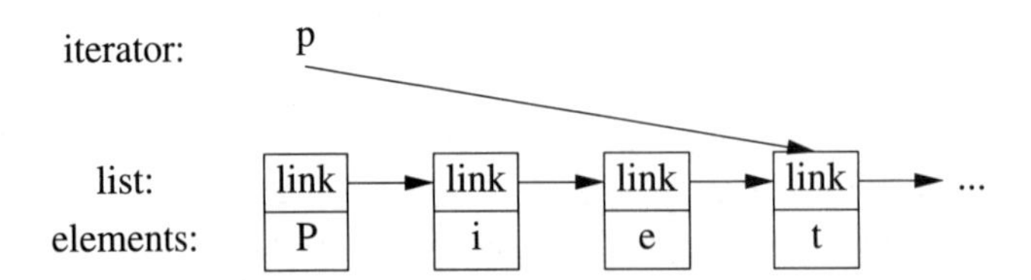

What is common for all iterators is their semantics and the naming of their operations. For example, applying ++ to any iterator yields an iterator that refers to the next element. Similarly, * yields the element to which the iterator refers. In fact, any object that obeys a few simple rules like these is an iterator. *Iterator* is a general idea, a concept (§8.2), and different kinds of iterators are made available as standard-library **concepts**, such as **forward_iterator** and **random_access_iterator** (§14.5). Furthermore, users rarely need to know the type of a specific iterator; each container "knows" its iterator types and makes them available under the conventional names **iterator** and **const_iterator**. For example, **list<Entry>::iterator** is the general iterator type for **list<Entry>**. We rarely have to worry about the details of how that type is defined.

In some cases, an iterator is not a member type, so the standard library offers **iterator_t<X>** that works wherever **X**'s iterator is defined.

13.3.1 Stream Iterators（流迭代器）

Iterators are general and useful concepts for dealing with sequences of elements in containers. However, containers are not the only place where we find sequences of elements. For example, an input stream produces a sequence of values, and we write a sequence of values to an output stream. Consequently, the notion of iterators can be usefully applied to input and output.

To make an **ostream_iterator**, we need to specify which stream will be used and the type of objects written to it. For example:

```
ostream_iterator<string> oo {cout};      // write strings to cout
```

The effect of assigning to ***oo** is to write the assigned value to **cout**. For example:

```
int main()
{
     *oo = "Hello, ";      // meaning cout<<"Hello, "
     ++oo;
     *oo = "world!\n";     // meaning cout<<"world!\n"
}
```

This is yet another way of writing the canonical message to standard output. The **++oo** is done to mimic writing into an array through a pointer. That way, we can use algorithms on streams. For example:

```
vector<string> v{ "Hello", ", ", "World!\n" };
copy(v, oo);
```

Similarly, an **istream_iterator** is something that allows us to treat an input stream as a read-only container. Again, we must specify the stream to be used and the type of values expected:

```
istream_iterator<string> ii {cin};
```

Input iterators are used in pairs representing a sequence, so we must provide an **istream_iterator** to indicate the end of input. This is the default **istream_iterator**:

```
istream_iterator<string> eos {};
```

Typically, **istream_iterator**s and **ostream_iterator**s are not used directly. Instead, they are provided as arguments to algorithms. For example, we can write a simple program to read a file, sort the words read, eliminate duplicates, and write the result to another file:

```
int main()
{
        string from, to;
        cin >> from >> to;                              // get source and target file names

        ifstream is {from};                             // input stream for file "from"
        istream_iterator<string> ii {is};               // input iterator for stream
        istream_iterator<string> eos {};                // input sentinel

        ofstream os {to};                               // output stream for file "to"
        ostream_iterator<string> oo {os,"\n"};          // output iterator for stream plus a separator

        vector<string> b {ii,eos};                      // b is a vector initialized from input
        sort(b);                                        // sort the buffer

        unique_copy(b,oo);                              // copy the buffer to output, discard replicated values

        return !is.eof() || !os;                        // return error state (§1.2.1, §11.4)
}
```

I used the range versions of **sort()** and **unique_copy()**. I could have used iterators directly, e.g., **sort(b.begin(),b.end())**, as is common in older code.

Please remember that to use both a traditional iterator version of a standard-library algorithm and its ranges counterpart, we need to either explicitly qualify the call of the range version or use a **using**-declaration (§9.3.2):

```
copy(v, oo);                    // potentially ambiguous
ranges::copy(v, oo);            // OK
using ranges::copy(v, oo);      // copy(v, oo) OK from here on
copy(v, oo);                    // OK
```

An **ifstream** is an **istream** that can be attached to a file (§11.7.2), and an **ofstream** is an **ostream** that can be attached to a file. The **ostream_iterator**'s second argument is used to delimit output values.

Actually, this program is longer than it needs to be. We read the strings into a **vector**, then we **sort()** them, and then we write them out, eliminating duplicates. A more elegant solution is not to store duplicates at all. This can be done by keeping the **strings** in a **set**, which does not keep duplicates and keeps its elements in order (§12.5). That way, we could replace the two lines using a **vector** with one using a **set** and replace **unique_copy()** with the simpler **copy()**:

```
set<string> b {ii,eos};         // collect strings from input
copy(b,oo);                     // copy buffer to output
```

We used the names **ii**, **eos**, and **oo** only once, so we could further reduce the size of the program:

```
int main()
{
        string from, to;
        cin >> from >> to;                      // get source and target file names
```

```
    ifstream is {from};                          // input stream for file "from"
    ofstream os {to};                            // output stream for file "to"

    set<string> b {istream_iterator<string>{is},istream_iterator<string>{}};     // read input
    copy(b,ostream_iterator<string>{os,"\n"});                                   // copy to output

    return !is.eof() || !os;                     // return error state (§1.2.1, §11.4)
}
```

It is a matter of taste and experience whether or not this last simplification improves readability.

13.4 Use of Predicates（使用谓词）

In the examples so far, the algorithms have simply "built in" the action to be done for each element of a sequence. However, we often want to make that action a parameter to the algorithm. For example, the `find` algorithm (§13.2, §13.5) provides a convenient way of looking for a specific value. A more general variant looks for an element that fulfills a specified requirement, a *predicate*. For example, we might want to search a `map` for the first value larger than `42`. A `map` allows us to access its elements as a sequence of *(key,value)* pairs, so we can search a `map<string,int>`'s sequence for a `pair<const string,int>` where the `int` is greater than `42`:

```
void f(map<string,int>& m)
{
    auto p = find_if(m,Greater_than{42});
    // ...
}
```

Here, `Greater_than` is a function object (§7.3.2) holding the value (`42`) to be compared against a `map` entry of type `pair<string,int>`:

```
struct Greater_than {
    int val;
    Greater_than(int v) : val{v} { }
    bool operator()(const pair<string,int>& r) const { return r.second>val; }
};
```

Alternatively and equivalently, we could use a lambda expression (§7.3.2):

```
auto p = find_if(m, [](const auto& r) { return r.second>42; });
```

A predicate should not modify the elements to which it is applied.

13.5 Algorithm Overview（标准库算法概览）

A general definition of an algorithm is "a finite set of rules which gives a sequence of operations for solving a specific set of problems [and] has five important features: Finiteness ... Definiteness ... Input ... Output ... Effectiveness" [Knuth,1968,§1.1]. In the context of the C++ standard library, an algorithm is a function template operating on sequences of elements.

The standard library provides many dozens of algorithms. The algorithms are defined in namespace std and presented in the <algorithm> and <numeric> headers. These standard-library algorithms all take sequences as inputs. A half-open sequence from b to e is referred to as [b:e). Here are a few examples:

Selected Standard Algorithms <algorithm>	
f=for_each(b,e,f)	For each element x in [b:e) do f(x)
p=find(b,e,x)	p is the first p in [b:e) so that *p==x
p=find_if(b,e,f)	p is the first p in [b:e) so that f(*p)
n=count(b,e,x)	n is the number of elements *q in [b:e) so that *q==x
n=count_if(b,e,f)	n is the number of elements *q in [b:e) so that f(*q)
replace(b,e,v,v2)	Replace elements *q in [b:e) so that *q==v with v2
replace_if(b,e,f,v2)	Replace elements *q in [b:e) so that f(*q) with v2
p=copy(b,e,out)	Copy [b:e) to [out:p)
p=copy_if(b,e,out,f)	Copy elements *q from [b:e) so that f(*q) to [out:p)
p=move(b,e,out)	Move [b:e) to [out:p)
p=unique_copy(b,e,out)	Copy [b:e) to [out:p); don't copy adjacent duplicates
sort(b,e)	Sort elements of [b:e) using < as the sorting criterion
sort(b,e,f)	Sort elements of [b:e) using f as the sorting criterion
(p1,p2)=equal_range(b,e,v)	[p1:p2) is the subsequence of the sorted sequence [b:e) with the value v; basically a binary search for v
p=merge(b,e,b2,e2,out)	Merge two sorted sequences [b:e) and [b2:e2) into [out:p)
p=merge(b,e,b2,e2,out,f)	Merge two sorted sequences [b:e) and [b2:e2) into [out:p) using f as the comparison

For each algorithm taking a [b:e) range, <ranges> offers a version that takes a range. Please remember (§9.3.2) that to use both a traditional iterator version of a standard-library algorithm and its ranges counterpart, you need to either explicitly qualify the call or use a using-declaration .

These algorithms, and many more (e.g., §17.3), can be applied to elements of containers, strings, and built-in arrays.

Some algorithms, such as replace() and sort(), modify element values, but no algorithm adds or subtracts elements of a container. The reason is that a sequence does not identify the container that holds the elements of the sequence. To add or delete elements, you need something that knows about the container (e.g., a back_inserter; §13.1) or directly refers to the container itself (e.g., push_back() or erase(); §12.2).

Lambdas are very common as operations passed as arguments. For example:

```
vector<int> v = {0,1,2,3,4,5};
for_each(v,[](int& x){ x=x*x; });                              // v=={0,1,4,9,16,25}
for_each(v.begin(),v.begin()+3,[](int& x){ x=sqrt(x); });      // v=={0,1,2,9,16,25}
```

The standard-library algorithms tend to be more carefully designed, specified, and implemented than the average hand-crafted loop. Know them and use them in preference to code written in the bare language.

13.6 Parallel Algorithms（并行算法）

When the same task is to be done to many data items, we can execute it in parallel on each data item provided the computations on different data items are independent:

- *parallel execution*: tasks are done on multiple threads (often running on several processor cores)
- *vectorized execution*: tasks are done on a single thread using vectorization, also known as *SIMD* ("Single Instruction, Multiple Data").

The standard library offers support for both and we can be specific about wanting sequential execution; in **<execution>** in namespace **execution**, we find:

- **seq**: sequential execution
- **par**: parallel execution (if feasible)
- **unseq**: unsequenced (vectorized) execution (if feasible)
- **par_unseq**: parallel and/or unsequenced (vectorized) execution (if feasible).

Consider **std::sort()**:

```
sort(v.begin(),v.end());                    // sequential
sort(seq,v.begin(),v.end());                // sequential (same as the default)
sort(par,v.begin(),v.end());                // parallel
sort(par_unseq,v.begin(),v.end());          // parallel and/or vectorized
```

Whether it is worthwhile to parallelize and/or vectorize depends on the algorithm, the number of elements in the sequence, the hardware, and the utilization of that hardware by programs running on it. Consequently, the *execution policy indicators* are just hints. A compiler and/or run-time scheduler will decide how much concurrency to use. This is all nontrivial and the rule against making statements about efficiency without measurement is very important here.

Unfortunately, the range versions of the parallel algorithms are not yet in the standard, but if we need them, they are easy to define:

```
void sort(auto pol, random_access_range auto& r)
{
        sort(pol,r.begin(),r.end());
}
```

Most standard-library algorithms, including all in the table in §13.5 except **equal_range**, can be requested to be parallelized and vectorized using **par** and **par_unseq** as for **sort()**. Why not **equal_range()**? Because so far nobody has come up with a worthwhile parallel algorithm for that.

Many parallel algorithms are used primarily for numeric data; see §17.3.1.

When requesting parallel execution, be sure to avoid data races (§18.2) and deadlock (§18.3).

13.7 Advice（建议）

[1] An STL algorithm operates on one or more sequences; §13.1.
[2] An input sequence is half-open and defined by a pair of iterators; §13.1.
[3] You can define your own iterators to serve special needs; §13.1.
[4] Many algorithms can be applied to I/O streams; §13.3.1.

[5] When searching, an algorithm usually returns the end of the input sequence to indicate "not found"; §13.2.
[6] Algorithms do not directly add or subtract elements from their argument sequences; §13.2, §13.5.
[7] When writing a loop, consider whether it could be expressed as a general algorithm; §13.2.
[8] Use **using**-type-aliases to clean up messy notation; §13.2.
[9] Use predicates and other function objects to give standard algorithms a wider range of meanings; §13.4, §13.5.
[10] A predicate must not modify its argument; §13.4.
[11] Know your standard-library algorithms and prefer them to hand-crafted loops; §13.5.

14

Ranges

（范围）

The strongest arguments prove nothing
so long as the conclusions are not verified by experience.
– Roger Bacon

- Introduction
- Views
- Generators
- Pipelines
- Concepts Overview
  Type Concepts; Iterator Concepts; Range Concepts
- Advice

14.1 Introduction（引言）

The standard-library offers algorithms both constrained using concepts (Chapter 8) and unconstrained (for compatibility). The constrained (concept) versions are in **<ranges>** in namespace **ranges**. Naurally, I prefer the versions using concepts. A **range** is a generalization of the C++98 sequences defined by {**begin(),end()**} pairs; it specifies what it takes to be a sequence of elements. A **range** can be defined by

- A **{begin,end}** pair of iterators
- A **{begin,n}** pair, where **begin** is an iterator and **n** is the number of elements
- A **{begin,pred}** pair, where **begin** is an iterator and **pred** is a predicate; if **pred(p)** is **true** for the iterator **p**, we have reached the end of the range. This allows us to have infinite ranges and ranges that are generated "on the fly" (§14.3).

This **range** concept is what allows us to say **sort(v)** rather than **sort(v.begin(),v.end())** as we had to using the STL since 1994. We can do similarly for our own algorithms:

```
template<forward_range R>
    requires sortable<iterator_t<R>>
void my_sort(R& r)                    // modern, concept-constrained version of my_sort
{
    return my_sort(r.begin(),end());      // use the 1994-style sort
}
```

Ranges allows us a more direct expression of something like 99% of the common uses of algorithms. In addition to the notational advantage, ranges offer some opportunities for optimization and eliminate a class of "silly errors," such as sort(v1.begin(),v2.end()) and sort(v.end(),v.begin()). Yes, such errors have been seen "in the wild."

Naturally, there are different kinds of ranges corresponding to the different kinds of iterators. In particular, input_range, forward_range, bidirectional_range, random_access_range, and contiguous_range are represented as concepts (§14.5).

14.2 Views（视图）

A view is a way of looking at a range. For example:

```
void user(forward_range auto& r)
{
    filter_view v {r, [](int x) { return x%2; } }; // view (only) odd numbers from r

    cout << "odd numbers: "
    for (int x : v)
        cout << x << ' ';
}
```

When reading from a filter_view, we read from its range. If the value read matches the predicate, it is returned; otherwise, the filter_view tries again with the next element from the range.

Many ranges are infinite. Also, we often only want a few values. Consequently, there are views for taking only a few values from a range:

```
void user(forward_range auto& r)
{
    filter_view v{r, [](int x) { return x%2; } };      // view (only) odd numbers in r
    take_view tv {v, 100 };                             // view at most 100 element from v

    cout << "odd numbers: "
    for (int x : tv)
        cout << x << ' ';
}
```

We can avoid naming the take_view by using it directly:

```
for (int x : take_view{v, 3})
    cout << x << ' ';
```

Similarly for the filter_view:

```
for (int x : take_view{ filter_view { r, [](int x) { return x % 2; } }, 3 })
      cout << x << ' ';
```

Such nesting of views can quickly get a bit cryptic, so there is an alternative: pipelines (§14.4).

The standard library offers many views, also known as *range adaptors*:

Standard-library views (range adaptors) <ranges> v is a view; r is range; p is a predicate; n is an integer	
v=all_view{r}	v is all elements from r
v=filter_view{r,p}	v is elements from r that meets p
v=transform_view{r,f}	v is the results of calling f on each element from r
v=take_view{r,n}	v is at most n elements from r
v=take_while_view{r,p}	v is elements from r until one doesn't meet the p
v=drop_view{r,n}	v is elements from r starting with the n+1th element
v=drop_while_view{r,p}	v is elements from r starting with the first element that doesn't meet p
v=join_view{r}	v is a flattened version of r; the elements of r must be ranges
v=split_view(r,d)	v is a range of sub-ranges of r determined by the delimiter d; d must be an element or a range
v=common_view(r)	v is r described by a (begin:end) pair
v=reverse_view{r}	v is the elements of r in reverse order; r must have bidirectional access
v=views::elements<n>(r)	v is the range of the nth elements of the tuple elements of r
v=keys_view{r}}	v is the range of the first elements of the pair elements of r
v=values_view{r}	v is the range of the second elements of the pair elements of r
v=ref_view{r}	v is range of elements that are references to the elements of r

A view offers an interface that's very similar to that of a range, so in most cases we can use a view wherever we can use a range and in the same way. The key difference is that a view doesn't own its elements; it is not responsible for deleting the elements of its underlying range - that's the range's responsibility. On the other hand, a view must not outlive its range:

```
auto bad()
{
      vector v = {1, 2, 3, 4};
      return filter_view{v,odd};              // v will be destroyed before the view
}
```

Views are supposed to be cheap to copy, so we pass them by value.

I have used simple standard types to keep examples trivial, but of course, we can have views of our own user-defined types. For example:

```
struct Reading {
      int location {};
      int temperature {};
      int humidity {};
      int air_pressure {};
      // ...
};
```

```
int average_temp(vector<Reading> readings)
{
	if (readings.size()==0) throw No_readings{};
	double s = 0;
	for (int x: views::elements<1>(readings))		// look at just the temperatures
		s += x;
	return s/readings.size();
}
```

14.3 Generators（生成器）

Often, a range needs to be generated on the fly. The standard library provides a few simple *generators* (aka *factories*) for that:

Range factories <ranges> v is a view; x is of the element type T; is is an istream	
v=empty_view<T>{}	v is an empty range of type T elements (had it had any)
v=single_view{x}	v is a range of the one element x
v=iota_view{x}	v is a infinite range of elements: x, x+1, x+2, ... incrementing is done using ++
v=iota_view{x,y}	v is a range of n elements: x, x+1, ..., y−1 incrementing is done using ++
v=istream_view<T>{is}	v is the range obtained by calling >> for T on is

The iota_views are useful for generating simple sequences. For example:

```
for (int x : iota_view(42,52))	// 42 43 44 45 46 47 48 49 50 51
	cout << x << ' ';
```

The istream_view gives us a simple way of using istreams in range-for loops:

```
for (auto x : istream_view<complex<double>(cin))
	cout << x << '\n';
```

Like other views, a istream_view can be composed with other views:

```
auto cplx = istream_view<complex<double>>(cin);

for (auto x : transform_view(cplx, [](auto z){ return z*z;}))
	cout << x << '\n';
```

An input of 1 2 3 produces 1 4 9.

14.4 Pipelines（管道）

For each standard-library view (§14.2), the standard library provides a function that produces a filter; that is, an object that can be used as an argument to the filter operator |. For example, filter() yields a filter_view. This allows us to combine filters in a sequence rather than presenting them as a set of nested function calls.

```
void user(forward_range auto& r)
{
    auto odd = [](int x) { return x % 2; };

    for (int x : r | views::filter(odd) | views::take(3))
        cout << x << ' ';
}
```

An input range **2 4 6 8 20** produces **1 2 3**.

The pipeline style (using the Unix pipeline operator **|**) is widely regarded as more readable than nested function calls. The pipeline works left to right; that is **f|g** the result of **f** is passed to **g**, so **r|f|g** means **(g_filter(f_filter(r)))**. The initial **r** has to be a range or a generator.

These filter functions are in namespace **ranges::views**:

```
void user(forward_range auto& r)
{
    for (int x : r | views::filter([](int x) { return x % 2; } ) | views::take(3) )
        cout << x << ' ';
}
```

I find that using **views::** explicitly makes code quite readable, but of course we can shorten the code further:

```
void user(forward_range auto& r)
{
    using namespace views;

    auto odd = [](int x) { return x % 2; };

    for (int x : r | filter(odd) | take(3) )
        cout << x << ' ';
}
```

The implementation of views and pipelines involves some quite hair-raising template metaprogramming, so if you are concerned about performance make sure to measure whether your implementation delivers what you need. If not, there is always a conventional workaround:

```
void user(forward_range auto& r)
{
    int count = 0;
    for (int x : r)
        if (x % 2) {
            cout << x << ' ';
            if (++count == 3) return;
        }
}
```

However, here the logic of what's going on is obscured.

14.5 Concepts Overview（概念概述）

The standard library offers many useful concepts:

- Concepts defining properties of types (§14.5.1)
- Concepts defining iterators (§14.5.2)
- Concepts defining ranges (§14.5.3)

14.5.1 Type Concepts（类型概念）

The concepts related to properties of types and the relations among types reflect the variety of types. These concepts help simplify most templates.

Core language concepts <concepts> T and U are types	
same_as<T,U>	T is the same as U
derived_from<T,U>	T is derived from U
convertible_to<T,U>	A T can be converted to a U
common_reference_with<T,U>	T and U share a common reference type
common_with<T,U>	T and U share a common type
integral<T>	T is an integral type
signed_integral<T>	T is a signed integral type
unsigned_integral<T>	T is an unsigned integral type
floating_point<T>	T is a floating point type
assignable_from<T,U>	A U can be assigned to a T
swappable_with<T,U>	A T can be swapped with a U
swappable<T>	swappable_with<T,T>

Many algorithms should work with combinations of related types, e.g., expressions with a mixture of **int**s and **double**s. We use **common_with** to say whether such a mix is mathematically sound. If **common_with<X,Y>** is **true**, we can use **common_type_t<X,Y>** to compare a **X** with a **Y** by first converting both to **common_type_t<X,Y>**. For example:

```
common_type<string, const char*> s1 = some_fct()
common_type<string, const char*> s2 = some_other_fct();

if (s1<s2) {
      // ...
}
```

To specify a common type for a pair of types, we specialize **common_type_t** used in the definition of **common**. For example:

```
using common_type_t<Bigint,long> = Bigint;        // for a suitable definition of Bigint
```

Fortunately, we don't need to define a **common_type_t** specialization unless we want to use operations on mixes of types for which a library doesn't (yet) have suitable definitions.

The concepts related to comparison are strongly influenced by [Stepanov,2009].

Comparison concepts <concepts>	
equality_comparable_with<T,U>	A T and a U can be compared for equivalence using ==
equality_comparable<T>	equality_comparablewith<T,T>
totally_ordered_with<T,U>	A T and a U can be compared using <, <=, >, and >= yielding a total order
totally_ordered<T>	strict_totally_ordered_with<T,T>
three_way_comparable_with<T,U>	A T and a U can be compared using <=> yielding a consistent result
three_way_comparable<T>	three_way_comparable_with<T,T>

The use of both **equality_comparable_with** and **equality_comparable** shows a (so far) missed opportunity to overload concepts.

Curiously, there is no standard **boolean** concept. I often need it, so here is a version:

```
template<typename B>
concept Boolean =
    requires(B x, B y) {
        { x = true };
        { x = false };
        { x = (x == y) };
        { x = (x != y) };
        { x = !x };
        { x = (x = y) };
    };
```

When writing templates, we often need to classify types.

Object concepts <concepts>	
destructible<T>	A T can be destroyed and have its address taken with unary &
constructible_from<T,Args>	A T can be constructed from an argument list of type Args
default_initializable<T>	A T can be default constructed
move_constructible<T>	A T can be move constructed
copy_constructible<T>	A T can be copy constructed and move constructed
movable<T>	move_constructable<T>, assignable<T&,T>, and swapable<T>
copyable<T>	copy_constructable<T>, moveable<T>, and assignable<T, const T&>
semiregular<T>	copyable<T> and default_constructable<T>
regular<T>	semiregular<T> and equality_comparable<T>

The ideal for types is **regular**. A **regular** type works roughly like an **int** and simplifies much of our thinking about how to use a type (§8.2). The lack of default == for classes means that most classes start out as **semiregular** even though most could and should be **regular**.

Whenever we pass an operation as a constrained template argument, we need to specify how it can be called, and sometimes also what assumptions we make of their semantics.

Callable concepts <concepts>	
invocable<F,Args>	An F can be invoked with an argument list of type Args
regular_invocable<F,Args>	invocable<F,Args> and is equality preserving
predicate<F,Args>	A regular_invocable<F,Args> returning a bool
relation<F,T,U>	predicate<F,T,U>
equivalence_relation<F,T,U>	A relation<F,T,U> that provides an equivalence relation
strict_weak_order<F,T,U>	A relation<F,T,U> that provides strict weak ordering

A function f() is *equality preserving* if x==y implies that f(x)==f(y). An invocable and a regular_invocable differ only semantically. We can't (currently) represent that in code so the names simply express our intent.

Similarly, a relation and an equivalence_relation differ only semantically. An equivalence relation is reflexive, symmetric, and transitive.

A relation and a strict_weak_order differ only semantically. Strict weak ordering is what the standard library usually assumes for comparisons, such as <.

14.5.2 Iterator Concepts（迭代器概念）

The traditional standard algorithms access their data through iterators, so we need concepts to classify properties of iterator types.

Iterator concepts <iterators>	
input_or_output_iterator<I>	An I can be incremented (++) and dereferenced (*)
sentinel_for<S,I>	An S is a sentinel for an Iterator type; that is, S is a predicate on I's value type
sized_sentinel_for<S,I>	A sentinel S where the - operator can be applied to I
input_iterator<I>	An I is an input iterator; * can be used for reading only
output_iterator<I>	An I is an output iterator; * can be used for writing only
forward_iterator<I>	An I is a forward iterator, supporting multi-pass and ==
bidirectional_iterator<I>	A forward_iterator<I> supporting --
random_access_iterator<I>	A bidirectional_iterator<I> supporting +, -, +=, -=, and []
contiguous_iterator<I>	A random_access_iterator<I> for elements in contiguous memory
permutable<I>	A forward_iterator<I> supporting move and swap
mergeable<I1,I2,R,O>	Can merge sorted sequences defined by I1 and I2 into O using relation<R>?
sortable<I>	Can sort sequences defined by I using less?
sortable<I,R>	Can sort sequences defined by I using relation<R>?

mergeable and sortable are simplified relative to their definition in C++20.

The different kinds (categories) of iterators are used to select the best algorithm for a given set of arguments; see §8.2.2 and §16.4.1. For an example of an input_iterator, see §13.3.1.

The basic idea of a sentinel is that we can iterate over a range starting at an iterator until the predicate becomes true for an element. That way, an iterator p and a sentinel s define a range [p:s(*p)). For example, we could define a predicate for a sentinel for traversing a C-style string using a pointer as the iterator. Unfortunately, this requires some boilerplate because the idea is to present the predicate as something that can't be confused with an ordinary iterator but that you can compare the iterator used to traverse the range:

```
template<class Iter>
class Sentinel {
public:
        Sentinel(int ee) : end(ee) { }
        Sentinel() :end(0) {}         // Concept sentinel_for requires a default constructor

        friend bool operator==(const Iter& p, Sentinel s) { return (*p == s.end); }
        friend bool operator!=(const Iter& p, Sentinel s) { return !(p == s); }
private:
        iter_value_t<const char*> end;        // the sentinel value
};
```

The **friend** declarator allows us to define the == and != binary functions for comparing an iterator to a sentinel within the scope of the class.

We can check that this **Sentinel** meets the requirements of **sentinel_for** for a **const char***:

```
static_assert(sentinel_for<Sentinel<const char*>, const char*>); // check the Sentinel for C-style strings
```

Finally, we can write a rather peculiar version of the "Hello, World!" program:

```
const char aa[] = "Hello, World!\nBye for now\n";

ranges::for_each(aa, Sentinel<const char*>('\n'), [](const char x) { cout << x; });
```

Yes, this really writes **Hello, World!** not followed by a newline.

14.5.3 Range Concepts（范围概念）

The range concepts defines the properties of ranges.

Range concepts <ranges>	
range<R>	An R is a range with a begin iterator and a sentinel
sized_range<R>	An R is a range that knows its size in constant time
view<R>	An R is a range with constant time copy, move, and assignment
common_range<R>	An R is a range with identical iterator and sentinel types
input_range<R>	An R is a range whose iterator type satisfies input_iterator
output_range<R>	An R is a range whose iterator type satisfies output_iterator
forward_range<R>	An R is a range whose iterator type satisfies forward_iterator
bidirectional_range<R>	An R is a range whose iterator type satisfies bidirectional_iterator
random_access_range<R>	An R is a range whose iterator type satisfies random_access_iterator
contiguous_range<R>	An R is a range whose iterator type satisfies contiguous_iterator

There are a few more concepts in **<ranges>**, but this set is a good start. The primary use of these concepts is to enable overloading of implementations based on type properties of their inputs (§8.2.2).

14.6 Advice（建议）

[1] When the pair-of-iterators style becomes tedious, use a range algorithm; §13.1; §14.1.
[2] When using a range algorithm, remember to explicitly introduce its name; §13.3.1.
[3] Pipelines of operations on a range can be expressed using **views**, **generator**s, and **filter**s; §14.2, §14.3, §14.4.
[4] To end a range with a predicate, you need to define a sentinel; §14.5.
[5] Using **static_assert**, we can check that a specific type meets the requirements of a concept; §8.2.4.
[6] If you want a range algorithm and there isn't one in the standard, just write your own; §13.6.
[7] The ideal for types is **regular**; §14.5.
[8] Prefer standard-library concepts where they apply; §14.5.
[9] When requesting parallel execution, be sure to avoid data races (§18.2) and deadlock (§18.3); §13.6.

15

Pointers and Containers

（指针和容器）

Education is what, when, and why to do things.
Training is how to do it.
– Richard Hamming

- Introduction
- Pointers
  unique_ptr and **shared_ptr**; **span**
- Containers
  array; **bitset**; **pair**; **tuple**
- Alternatives
  variant; **optional**; **any**
- Advice

15.1 Introduction（引言）

C++ offers simple built-in low-level types to hold and refer to data: objects and arrays hold data; pointers and arrays refer to such data. However, we need to support both more specialized and more general ways for holding and using data. For example, the standard-library containers (Chapter 12) and iterators (§13.3) are designed to support general algorithms.

The main commonality among the container and pointer abstractions is that their correct and efficient use requires encapsulation of data together with a set of functions to access and manipulate them. For example, pointers are very general and efficient abstractions of machine addresses, but using them correctly to represent ownership of resources has proven excessively difficult. So, the standard-library offers resource-management pointers; that is, classes that encapsulate pointers and provide operations that simplify their correct use.

These standard-library abstractions encapsulate built-in language types and are required to perform as well in time and space as correct uses of those types.

There is nothing "magic" about these types. We can design and implement our own "smart pointers" and specialized containers as needed using the same techniques as are used for the standard-library ones.

15.2 Pointers（指针类型）

The general notion of a *pointer* is something that allows us to refer to an object and to access it according to its type. A built-in pointer, such as **int***, is an example but there are many more.

Pointers	
T*	A built-in pointer type: points to an object of type **T** or to a contiguously-allocated sequence of elements of type **T**
T&	A built-in reference type: refers to an object of type **T**; a pointer with implicit dereference (§1.7)
unique_ptr<T>	An owning pointer to a **T**
shared_ptr<T>	A pointer to an object of type **T**; ownership is shared among all **shared_ptr**'s to that **T**
weak_ptr<T>	A pointer to an object owned by a **shared_ptr**; must be converted to a **shared_ptr** to access the object
span<T>	A pointer to a contiguous sequence of **T**s (§15.2.2)
string_view<T>	A pointer to a **const** sub-string (§10.3)
X_iterator<C>	A sequence of elements from **C**; The **X** in the name indicates the kind of iterator (§13.3)

There can be more than one pointer pointing to an object. An *owning* pointer is one that is responsible for eventually deleting the object it refers to. A *non-owning* pointer (e.g., a **T*** or a **span**) can *dangle*; that is, point to a location where an object has been **deleted** or gone out of scope.

Reading or writing through a dangling pointer is one of the nastiest kinds of bugs. The result of doing so is technically undefined. In practice, that often means accessing an object that happens to occupy the location. Then, a read means getting an arbitrary value, and a write scrambles an unrelated data structure. The best we can hope for is a crash; that's usually preferable to a wrong result.

The C++ Core Guidelines [CG] offers rules for avoiding this and advice for statically checking that it never happens. However, here are a few approaches for avoiding pointer problems:

- Don't retain a pointer to a local object after the object goes out of scope. In particular, never return a pointer to a local object from a function or store a pointer of uncertain provenance in a long-lived data structure. Systematic use of containers and algorithms (Chapter 12, Chapter 13) often saves us from employing programming techniques that make it hard to avoid pointer problems.
- Use owning pointers to objects allocated on the free store.
- Pointers to static objects (e.g., global variables) can't dangle.
- Leave pointer arithmetic to the implementation of resource handles (such as **vector**s and **unordered_map**s).
- Remember that **string_view**s and **span**s are kinds of non-owning pointers.

15.2.1 unique_ptr and shared_ptr（unique_ptr及shared_ptr）

One of the key tasks of any nontrivial program is to manage resources. A resource is something that must be acquired and later (explicitly or implicitly) released. Examples are memory, locks, sockets, thread handles, and file handles. For a long-running program, failing to release a resource in a timely manner ("a leak") can cause serious performance degradation (§12.7) and possibly even a miserable crash. Even for short programs, a leak can become an embarrassment, say by causing a resource shortage increasing the run time by orders of magnitude.

The standard-library components are designed not to leak resources. To do this, they rely on the basic language support for resource management using constructor/destructor pairs to ensure that a resource doesn't outlive an object responsible for it. The use of a constructor/destructor pair in **Vector** to manage the lifetime of its elements is an example (§5.2.2) and all standard-library containers are implemented in similar ways. Importantly, this approach interacts correctly with error handling using exceptions. For example, this technique is used for the standard-library lock classes:

```
mutex m; // used to protect access to shared data

void f()
{
        scoped_lock lck {m};        // acquire the mutex m
        // ... manipulate shared data ...
}
```

A **thread** will not proceed until **lck**'s constructor has acquired the **mutex** (§18.3). The corresponding destructor releases the **mutex**. So, in this example, **scoped_lock**'s destructor releases the **mutex** when the thread of control leaves **f()** (through a **return**, by "falling off the end of the function," or through an exception throw).

This is an application of RAII (the "Resource Acquisition Is Initialization" technique; §5.2.2). RAII is fundamental to the idiomatic handling of resources in C++. Containers (such as **vector** and **map**, **string**, and **iostream**) manage their resources (such as file handles and buffers) similarly.

The examples so far take care of objects defined in a scope, releasing the resources they acquire at the exit from the scope, but what about objects allocated on the free store? In **<memory>**, the standard library provides two "smart pointers" to help manage objects on the free store:

- **unique_ptr** represents unique ownership (its destructor destroys its object)
- **shared_ptr** represents shared ownership (the last shared pointer's destructor destroys the object)

The most basic use of these "smart pointers" is to prevent memory leaks caused by careless programming. For example:

```
void f(int i, int j)        // X* vs. unique_ptr<X>
{
        X* p = new X;                        // allocate a new X
        unique_ptr<X> sp {new X};            // allocate a new X and give its pointer to unique_ptr
        // ...
```

```
        if (i<99) throw Z{};          // may throw an exception
        if (j<77) return;             // may return "early"
        // ... use p and sp ..
        delete p;                     // destroy *p
}
```

Here, we "forgot" to delete **p** if **i<99** or if **j<77**. On the other hand, **unique_ptr** ensures that its object is properly destroyed whichever way we exit **f()** (by throwing an exception, by executing **return**, or by "falling off the end"). Ironically, we could have solved the problem simply by *not* using a pointer and *not* using **new**:

```
void f(int i, int j)      // use a local variable
{
        X x;
        // ...
}
```

Unfortunately, overuse of **new** (and of pointers and references) seems to be an increasing problem.

However, when you really need the semantics of pointers, **unique_ptr** is a lightweight mechanism with no space or time overhead compared to correct use of a built-in pointer. Its further uses include passing free-store allocated objects in and out of functions:

```
unique_ptr<X> make_X(int i)
        // make an X and immediately give it to a unique_ptr
{
        // ... check i, etc. ...
        return unique_ptr<X>{new X{i}};
}
```

A **unique_ptr** is a handle to an individual object (or an array) in much the same way that a **vector** is a handle to a sequence of objects. Both control the lifetime of other objects (using RAII) and both rely on elimination of copying or on move semantics to make **return** simple and efficient (§6.2.2).

The **shared_ptr** is similar to **unique_ptr** except that **shared_ptr**s are copied rather than moved. The **shared_ptr**s for an object share ownership of an object; that object is destroyed when the last of its **shared_ptr**s is destroyed. For example:

```
void f(shared_ptr<fstream>);
void g(shared_ptr<fstream>);

void user(const string& name, ios_base::openmode mode)
{
        shared_ptr<fstream> fp {new fstream(name,mode)};
        if (!*fp)                     // make sure the file was properly opened
                throw No_file{};

        f(fp);
        g(fp);
        // ...
}
```

Now, the file opened by fp's constructor will be closed by the last function to (explicitly or implicitly) destroy a copy of fp. Note that f() or g() may spawn a task holding a copy of fp or in some other way store a copy that outlives user(). Thus, shared_ptr provides a form of garbage collection that respects the destructor-based resource management of the memory-managed objects. This is neither cost free nor exorbitantly expensive, but it does make the lifetime of the shared object hard to predict. Use shared_ptr only if you actually need shared ownership.

Creating an object on the free store and then passing the pointer to it to a smart pointer is a bit verbose. It also allows for mistakes, such as forgetting to pass a pointer to a unique_ptr or giving a pointer to something that is not on the free store to a shared_ptr. To avoid such problems, the standard library (in <memory>) provides functions for constructing an object and returning an appropriate smart pointer, make_shared() and make_unique(). For example:

```
struct S {
      int i;
      string s;
      double d;
      // ...
};

auto p1 = make_shared<S>(1,"Ankh Morpork",4.65);      // p1 is a shared_ptr<S>
auto p2 = make_unique<S>(2,"Oz",7.62);                // p2 is a unique_ptr<S>
```

Now, p2 is a unique_ptr<S> pointing to a free-store-allocated object of type S with the value {2,"Oz"s,7.62}.

Using make_shared() is not just more convenient than separately making an object using new and then passing it to a shared_ptr – it is also notably more efficient because it does not need a separate allocation for the use count that is essential in the implementation of a shared_ptr.

Given unique_ptr and shared_ptr, we can implement a complete "no naked new" policy (§5.2.2) for many programs. However, these "smart pointers" are still conceptually pointers and therefore only my second choice for resource management – after containers and other types that manage their resources at a higher conceptual level. In particular, shared_ptrs do not in themselves provide any rules for which of their owners can read and/or write the shared object. Data races (§18.5) and other forms of confusion are not addressed simply by eliminating the resource management issues.

When do we use "smart pointers" (such as unique_ptr) rather than resource handles with operations designed specifically for the resource (such as vector or thread)? Unsurprisingly, the answer is "when we need pointer semantics."

- When we share an object, we need pointers (or references) to refer to the shared object, so a shared_ptr becomes the obvious choice (unless there is an obvious single owner).
- When we refer to a polymorphic object in classical object-oriented code (§5.5), we need a pointer (or a reference) because we don't know the exact type of the object referred to (or even its size), so a unique_ptr becomes the obvious choice.
- A shared polymorphic object typically requires shared_ptrs.

We do *not* need to use a pointer to return a collection of objects from a function; a container that is a resource handle will do that simply and efficiently by relying on copy elision (§3.4.2) and move semantics (§6.2.2).

15.2.2 span

Traditionally, range errors have been a major source of serious errors in C and C++ programs, leading to wrong results, crashes, and security problems. The use of containers (Chapter 12), algorithms (Chapter 13), and range-**for** has significantly reduced this problem, but more can be done. A key source of range errors is that people pass pointers (raw or smart) and then rely on convention to know the number of elements pointed to. The best advice for code outside resource handles is to assume that at most one object is pointed to [CG: F.22], but without support that advice is unmanageable. The standard-library **string_view** (§10.3) can help, but that is read-only and for characters only. Most programmers need more. For example, when writing into and reading out of buffers in lower-level software, it is notoriously difficult to maintain high performance while still avoiding range errors ("buffer overruns"). A **span** from **<span>** is basically a (pointer,length) pair denoting a sequence of elements:

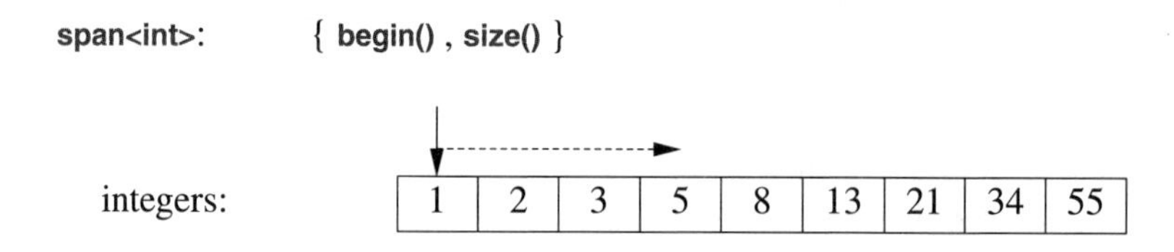

A **span** gives access to a contiguous sequence of elements. The elements can be stored in many ways, including in **vector**s and built-in arrays. Like a pointer, a **span** does not own the characters to which it points. In that, it resembles a **string_view** (§10.3) and an STL pair of iterators (§13.3).

Consider a common interface style:

```
void fpn(int* p, int n)
{
        for (int i = 0; i<n; ++i)
                p[i] = 0;
}
```

We assume that **p** points to **n** integers. Unfortunately, this assumption is simply a convention, so we can't use it to write a range-**for** loop and the compiler cannot implement cheap and effective range checking. Also, our assumption can be wrong:

```
void use(int x)
{
        int a[100];
        fpn(a,100);             // OK
        fpn(a,1000);            // oops, my finger slipped! (range error in fpn)
        fpn(a+10,100);          // range error in fpn
        fpn(a,x);               // suspect, but looks innocent
}
```

We can do better using a **span**:

```
void fs(span<int> p)
{
    for (int& x : p)
        x = 0;
}
```

We can use **fs** like this:

```
void use(int x)
{
    int a[100];
    fs(a);                  // implicitly creates a span<int>{a,100}
    fs(a,1000);             // error: span expected
    fs({a+10,100});         // a range error in fs
    fs({a,x});              // obviously suspect
}
```

That is, the common case, creating a **span** directly from an array, is now safe (the compiler computes the element count) and notationally simple. In other cases, the probability of mistakes is lowered and error-detection is made easier because the programmer has to explicitly compose a **span**.

The common case where a **span** is passed along from function to function is simpler than for (pointer,count) interfaces and obviously doesn't require extra checking:

```
void f1(span<int> p);

void f2(span<int> p)
{
    // ...
    f1(p);
}
```

As for containers, when **span** is used for subscripting (e.g., **r[i]**), range checking is not done and an out-of-range access is undefined behavior. Naturally, an implementation can implement that undefined behavior as range checking, but sadly few do. The original **gsl::span** from the Core Guidelines support library [CG] does range checking.

15.3 Containers（容器）

The standard provides several containers that don't fit perfectly into the STL framework (Chapter 12, Chapter 13). Examples are built-in arrays, **array**, and **string**. I sometimes refer to those as "almost containers," but that is not quite fair: they hold elements, so they are containers, but each has restrictions or added facilities that make them awkward in the context of the STL. Describing them separately also simplifies the description of the STL.

Containers	
T[N]	Built-in array: a fixed-size contiguously allocated sequence of N elements of type T; implicitly converts to a T*
array<T,N>	A fixed-size contiguously allocated sequence of N elements of type T; like the built-in array, but with most problems solved
bitset<N>	A fixed-size sequence of N bits
vector<bool>	A sequence of bits compactly stored in a specialization of vector
pair<T,U>	Two elements of types T and U
tuple<T...>	A sequence of an arbitrary number of elements of arbitrary types
basic_string<C>	A sequence of characters of type C; provides string operations
valarray<T>	An array of numeric values of type T; provides numeric operations

Why does the standard provide so many containers? They serve common but different (often overlapping) needs. If the standard library didn't provide them, many people would have to design and implement their own. For example:

- **pair** and **tuple** are heterogeneous; all other containers are homogeneous (all elements are of the same type).
- **array**, and **tuple** elements are contiguously allocated; **list** and **map** are linked structures.
- **bitset** and **vector<bool>** hold bits and access them through proxy objects; all other standard-library containers can hold a variety of types and access elements directly.
- **basic_string** requires its elements to be some form of character and to provide string manipulation, such as concatenation and locale-sensitive operations.
- **valarray** requires its elements to be numbers and to provide numerical operations.

All of these containers can be seen as providing specialized services needed by large communities of programmers. No single container could serve all of these needs because some needs are contradictory, for example, "ability to grow" vs. "guaranteed to be allocated in a fixed location," and "elements do not move when elements are added" vs. "contiguously allocated."

15.3.1 array

An **array**, defined in **<array>**, is a fixed-size sequence of elements of a given type where the number of elements is specified at compile time. Thus, an **array** can be allocated with its elements on the stack, in an object, or in static storage. The elements are allocated in the scope where the **array** is defined. An **array** is best understood as a built-in array with its size firmly attached, without implicit, potentially surprising conversions to pointer types, and with a few convenience functions provided. There is no overhead (time or space) involved in using an **array** compared to using a built-in array. An **array** does *not* follow the "handle to elements" model of STL containers. Instead, an **array** directly contains its elements. It is nothing more or less than a safer version of a built-in array.

This implies that an **array** can and must be initialized by an initializer list:

```
array<int,3> a1 = {1,2,3};
```

The number of elements in the initializer must be equal to or less than the number of elements specified for the **array**.

The element count is not optional, the element count must be a constant expression, the number of elements must be positive, and the element type must be explicitly stated:

```
void f(int n)
{
      array<int> a0 = {1,2,3};                          // error size not specified
      array<string,n> a1 = {"John's", "Queens' "};  // error: size not a constant expression
      array<string,0> a2;                               // error: size must be positive
      array<2> a3 = {"John's", "Queens' "};           // error: element type not stated
      // ...
}
```

If you need the element count to be a variable, use **vector**.

When necessary, an **array** can be explicitly passed to a C-style function that expects a pointer. For example:

```
void f(int* p, int sz);     // C-style interface

void g()
{
      array<int,10> a;

      f(a,a.size());              // error: no conversion
      f(a.data(),a.size());       // C-style use

      auto p = find(a,777);       // C++/STL-style use (a range is passed)
      // ...
}
```

Why would we use an **array** when **vector** is so much more flexible? An **array** is less flexible so it is simpler. Occasionally, there is a significant performance advantage to be had by directly accessing elements allocated on the stack rather than allocating elements on the free store, accessing them indirectly through the **vector** (a handle), and then deallocating them. On the other hand, the stack is a limited resource (especially on some embedded systems), and stack overflow is nasty. Also, there are application areas, such as safety-critical real-time control, where free store allocation is banned. For example, use of **delete** may lead to fragmentation (§12.7) or memory exhaustion (§4.3).

Why would we use an **array** when we could use a built-in array? An **array** knows its size, so it is easy to use with standard-library algorithms, and it can be copied using **=**. For example:

```
array<int,3> a1 = {1, 2, 3 };
auto a2 = a1;   // copy
a2[1] = 5;
a1 = a2;        // assign
```

However, my main reason to prefer **array** is that it saves me from surprising and nasty conversions to pointers. Consider an example involving a class hierarchy:

```
void h()
{
    Circle a1[10];
    array<Circle,10> a2;
    // ...
    Shape* p1 = a1;    // OK: disaster waiting to happen
    Shape* p2 = a2;    // error: no conversion of array<Circle,10> to Shape* (Good!)
    p1[3].draw();      // disaster
}
```

The "disaster" comment assumes that **sizeof(Shape)<sizeof(Circle)**, so subscripting a **Circle[]** through a **Shape*** gives a wrong offset. All standard containers provide this advantage over built-in arrays.

15.3.2 bitset

Aspects of a system, such as the state of an input stream, are often represented as a set of flags indicating binary conditions such as good/bad, true/false, and on/off. C++ supports the notion of small sets of flags efficiently through bitwise operations on integers (§1.4). Class **bitset<N>** generalizes this notion by providing operations on a sequence of **N** bits **[0:N)**, where **N** is known at compile time. For sets of bits that don't fit into a **long long int** (often 64 bits), using a **bitset** is much more convenient than using integers directly. For smaller sets, **bitset** is usually optimized. If you want to name the bits, rather than numbering them, you can use a **set** (§12.5) or an enumeration (§2.4).

A **bitset** can be initialized with an integer or a string:

```
bitset<9> bs1 {"110001111"};
bitset<9> bs2 {0b1'1000'1111};    // binary literal using digit separators (§1.4)
```

The usual bitwise operators (§1.4) and the left- and right-shift operators (**<<** and **>>**) can be applied:

```
bitset<9> bs3 = ~bs1;        // complement: bs3=="001110000"
bitset<9> bs4 = bs1&bs3;     // all zeros
bitset<9> bs5 = bs1<<2;      // shift left: bs5 = "000111100"
```

The shift operators (here, **<<**) "shift in" zeros.

The operations **to_ullong()** and **to_string()** provide the inverse operations to the constructors. For example, we could write out the binary representation of an **int**:

```
void binary(int i)
{
    bitset<8*sizeof(int)> b = i;        // assume 8-bit byte (see also §17.7)
    cout << b.to_string() << '\n';      // write out the bits of i
}
```

This prints the bits represented as **1**s and **0**s from left to right, with the most significant bit leftmost, so that argument **123** would give the output

```
00000000000000000000000001111011
```

For this example, it is simpler to directly use the **bitset** output operator:

```
void binary2(int i)
{
      bitset<8*sizeof(int)> b = i;        // assume 8-bit byte (see also §17.7)
      cout << b << '\n';                   // write out the bits of i
}
```

A **bitset** offers many functions for using and manipulating sets of bits, such as **all()**, **any()**, **none()**, **count()**, **flip()**.

15.3.3 pair

It is fairly common for a function to return two values. There are many ways of doing that, the simplest and often the best is to define a **struct** for the purpose. For example, we can return a value and a success indicator:

```
struct My_res {
      Entry* ptr;
      Error_code err;
};

My_res complex_search(vector<Entry>& v, const string& s)
{
      Entry* found = nullptr;
      Error_code err = Error_code::found;
      // ... search for s in v ...
      return {found,err};
}

void user(const string& s)
{
      My_res r = complex_search(entry_table,s);     // search entry_table
      if (r.err != Error_code::good) {
            // ... handle error ...
      }
      // ... use r.ptr ....
}
```

We could argue that encoding failure as the end iterator or a **nullptr** is more elegant, but that can express just one kind of failure. Often, we would like to return two separate values. Defining a specific named **struct** for each pair of values often works well and is quite readable if the names of the "pair of values" **struct**s and their members are well chosen. However, for large code bases it can lead to a proliferation of names and conventions, and it doesn't work well for generic code where consistent naming is essential. Consequently, the standard library provides **pair** as a general support for the "pair of values" use cases. Using **pair**, our simple example becomes:

```
pair<Entry*,Error_code> complex_search(vector<Entry>& v, const string& s)
{
    Entry* found = nullptr;
    Error_code err = Error_code::found;
    // ... search for s in v ...
    return {found,err};
}

void user(const string& s)
{
    auto r = complex_search(entry_table,s);      // search entry_table
    if (r.second != Error_code::good) {
        // ... handle error ...
    }
    // ... use r.first ....
}
```

The members of **pair** are named **first** and **second**. That makes sense from an implementer's point of view, but in application code we may want to use our own names. Structured binding (§3.4.5) can be used to deal with that:

```
void user(const string& s)
{
    auto [ptr,success] = complex_search(entry_table,s);     // search entry_table
    if (success != Error_code::good)
        // ... handle error ...
    }
    // ... use r.ptr ....
}
```

The standard-library **pair** (from **<utility>**) is quite frequently used for "pair of values" use cases in the standard library and elsewhere. For example, the standard-library algorithm **equal_range** returns a **pair** of iterators specifying a subsequence meeting a predicate:

```
template<typename Forward_iterator, typename T, typename Compare>
    pair<Forward_iterator,Forward_iterator>
    equal_range(Forward_iterator first, Forward_iterator last, const T& val, Compare cmp);
```

Given a sorted sequence [**first:last**), **equal_range()** will return the **pair** representing the subsequence that matches the predicate **cmp**. We can use that to search in a sorted sequence of **Records**:

```
auto less = [](const Record& r1, const Record& r2) { return r1.name<r2.name;};     // compare names

void f(const vector<Record>& v)          // assume that v is sorted on its "name" field
{
    auto [first,last] = equal_range(v.begin(),v.end(),Record{"Reg"},less);

    for (auto p = first; p!=last; ++p)           // print all equal records
        cout << *p;                              // assume that << is defined for Record
}
```

A **pair** provides operators, such as =, ==, and <, if its elements do. Type deduction makes it easy to create a **pair** without explicitly mentioning its type. For example:

```
void f(vector<string>& v)
{
      pair p1 {v.begin(),2};                 // one way
      auto p2 = make_pair(v.begin(),2);      // another way
      // ...
}
```

Both **p1** and **p2** are of type **pair<vector<string>::iterator,int>**.

When code doesn't need to be generic, a simple struct with named members often leads to more maintainable code.

15.3.4 tuple

Like arrays, the standard-library containers are homogeneous; that is, all their elements are of a single type. However, sometimes we want to treat a sequence of elements of different types as a single object; that is, we want a heterogeneous container; **pair** is an example, but not all such heterogeneous sequences have just two elements. The standard library provides **tuple** as a generalization of **pair** with zero or more elements:

```
tuple t0 {};                                        // empty
tuple<string,int,double> t1 {"Shark",123,3.14};     // the type is explicitly specified
auto t2 = make_tuple(string{"Herring"},10,1.23);    // the type is deduced to tuple<string,int,double>
tuple t3 {"Cod"s,20,9.99};                          // the type is deduced to tuple<string,int,double>
```

The elements (members) of a **tuple** are independent; there is no invariant (§4.3) maintained among them. If we want an invariant, we must encapsulate the **tuple** in a class that enforces it.

For a single, specific use, a simple **struct** is often ideal, but there are many generic uses where the flexibility of **tuple** saves us from having to define many **struct**s at the cost of not having mnemonic names for the members. Members of a **tuple** are accessed through a **get** function template. For example:

```
string fish = get<0>(t1);          // get the first element: "Shark"
int count = get<1>(t1);            // get the second element: 123
double price = get<2>(t1);         // get the third element: 3.14
```

The elements of a **tuple** are numbered (starting with zero) and the index argument to **get()** must be a constant. The function **get** is a template function taking the index as a template value argument (§7.2.2).

Accessing members of a **tuple** by their index is general, ugly, and somewhat error-prone. Fortunately, an element of a **tuple** with a unique type in that **tuple** can be "named" by its type:

```
auto fish = get<string>(t1);       // get the string: "Shark"
auto count = get<int>(t1);         // get the int: 123
auto price = get<double>(t1);      // get the double: 3.14
```

We can use **get<>** for writing also:

```
get<string>(t1) = "Tuna";          // write to the string
get<int>(t1) = 7;                  // write to the int
get<double>(t1) = 312;             // write to the double
```

Most uses of **tuples** are hidden in implementations of higher-level constructs. For example, we could access the members of **t1** using structured binding (§3.4.5):

```
auto [fish, count, price] = t1;
cout << fish << ' ' << count << ' ' << price << '\n';    // read
fish = "Sea Bass";                                        // write
```

Typically, such binding and its underlying use of a **tuple** is used for a function call:

```
auto [fish, count, price] = todays_catch();
cout << fish << ' ' << count << ' ' << price << '\n';
```

The real strength of **tuple** is when you have to store or pass around an unknown number of elements of unknown types as an object.

Explicitly, iterating over the elements of a **tuple** is a bit messy, requiring recursion and compile-time evaluation of the function body:

```
template <size_t N = 0, typename... Ts>
constexpr void print(tuple<Ts...> tup)
{
        if constexpr (N<sizeof...(Ts)) {        // not yet at the end?
                cout << get<N>(tup) << ' ';     // print the Nth element
                print<N+1>(tup);                // print the next element
        }
}
```

Here, **sizeof...(Ts)** gives the number of elements in **Ts**.

Using **print()** is straightforward:

```
print(t0);          // no output
print(t2);          // Herring 10 1.23
print(tuple{ "Norah", 17, "Gavin", 14, "Anya", 9, "Courtney", 9, "Ada", 0 });
```

Like **pair**, **tuple** provides operators, such as **=**, **==**, and **<**, if its elements do. There are also conversions between a **pair** and a **tuple** with two members,

15.4 Alternatives（可变类型容器）

The standard offers three types to express alternatives:

Alternatives	
union	A built-in type that holds one of a set of alternatives (§2.5)
variant<T...>	One of a specified set of alternatives (in **<variant>**)
optional<T>	A value of type **T** or no value (in **<optional>**)
any	A value one of an unbounded set of alternative types (in **<any>**)

These types offer related functionality to a user. Unfortunately, they don't offer a unified interface.

15.4.1 variant

A variant<A,B,C> is often a safer and more convenient alternative to explicitly using a union (§2.5). Possibly the simplest example is to return either a value or an error code:

```
variant<string,Error_code> compose_message(istream& s)
{
      string mess;
      // ... read from s and compose message ...
      if (no_problems)
            return mess;                              // return a string
      else
            return Error_code{some_problem};          // return an Error_code
}
```

When you assign or initialize a variant with a value, it remembers the type of that value. Later, we can inquire what type the variant holds and extract the value. For example:

```
auto m = compose_message(cin);

if (holds_alternative<string>(m)) {
      cout << get<string>(m);
}
else {
      auto err = get<Error_code>(m);
      // ... handle error ...
}
```

This style appeals to some people who dislike exceptions (see §4.4), but there are more interesting uses. For example, a simple compiler may need to distinguish between different kinds of nodes with different representations:

```
using Node = variant<Expression,Statement,Declaration,Type>;

void check(Node* p)
{
      if (holds_alternative<Expression>(*p)) {
            Expression& e = get<Expression>(*p);
            // ...
      }
      else if (holds_alternative<Statement>(*p)) {
            Statement& s = get<Statement>(*p);
            // ...
      }
      // ... Declaration and Type ...
}
```

This pattern of checking alternatives to decide on the appropriate action is so common and relatively inefficient that it deserves direct support:

```
void check(Node* p)
{
    visit(overloaded {
        [](Expression& e) { /* ... */ },
        [](Statement& s) { /* ... */ },
        // ... Declaration and Type ...
    }, *p);
}
```

This is basically equivalent to a virtual function call, but potentially faster. As with all claims of performance, this "potentially faster" should be verified by measurements when performance is critical. For most uses, the difference in performance is insignificant.

The **overloaded** class is necessary and strangely enough, not standard. It's a "piece of magic" that builds an overload set from a set of arguments (usually lambdas):

```
template<class... Ts>
struct overloaded : Ts... {        // variadic template (§8.4)
    using Ts::operator()...;
};

template<class... Ts>
    overloaded(Ts...) -> overloaded<Ts...>;   // deduction guide
```

The "visitor" **visit** then applies **()** to the **overload** object, which selects the most appropriate lambda to call according to the overload rules.

A *deduction guide* is a mechanism for resolving subtle ambiguities, primarily for constructors of class templates in foundation libraries (§7.2.3).

If we try to access a **variant** holding a different type from the expected one, **bad_variant_access** is thrown.

15.4.2 optional

An **optional<A>** can be seen as a special kind of **variant** (like a **variant<A,nothing>**) or as a generalization of the idea of an **A*** either pointing to an object or being **nullptr**.

An **optional** can be useful for functions that may or may not return an object:

```
optional<string> compose_message(istream& s)
{
    string mess;

    // ... read from s and compose message ...

    if (no_problems)
        return mess;
    return {};          // the empty optional
}
```

Given that, we can write

```
if (auto m = compose_message(cin))
      cout << *m;            // note the dereference (*)
else {
      // ... handle error ...
}
```

This appeals to some people who dislike exceptions (see §4.4). Note the curious use of *. An **optional** is treated as a pointer to its object rather than the object itself.

The **optional** equivalent to **nullptr** is the empty object, {}. For example:

```
int sum(optional<int> a, optional<int> b)
{
      int res = 0;
      if (a) res+=*a;
      if (b) res+=*b;
      return res;
}

int x = sum(17,19);          // 36
int y = sum(17,{});          // 17
int z = sum({},{});          // 0
```

If we try to access an **optional** that does not hold a value, the result is undefined; an exception is *not* thrown. Thus, **optional** is not guaranteed type safe. Don't try:

```
int sum2(optional<int> a, optional<int> b)
{
      return *a+*b;   // asking for trouble
}
```

15.4.3 any

An **any** can hold an arbitrary type and know which type (if any) it holds. It is basically an unconstrained version of **variant**:

```
any compose_message(istream& s)
{
      string mess;

      // ... read from s and compose message ...

      if (no_problems)
            return mess;                // return a string
      else
            return error_number;        // return an int
}
```

When you assign or initialize an **any** with a value, it remembers the type of that value. Later, we can extract the value held by the **any** by asserting the value's expected type. For example:

```
auto m = compose_message(cin);
string& s = any_cast<string>(m);
cout << s;
```

If we try to access an **any** holding a different type than the expected one, **bad_any_access** is thrown.

15.5 Advice（建议）

[1] A library doesn't have to be large or complicated to be useful; §16.1.
[2] A resource is anything that has to be acquired and (explicitly or implicitly) released; §15.2.1.
[3] Use resource handles to manage resources (RAII); §15.2.1; [CG: R.1].
[4] The problem with a **T*** is that it can be used to represent anything, so we cannot easily determine a "raw" pointer's purpose; §15.2.1.
[5] Use **unique_ptr** to refer to objects of polymorphic type; §15.2.1; [CG: R.20].
[6] Use **shared_ptr** to refer to shared objects (only); §15.2.1; [CG: R.20].
[7] Prefer resource handles with specific semantics to smart pointers; §15.2.1.
[8] Don't use a smart pointer where a local variable will do; §15.2.1.
[9] Prefer **unique_ptr** to **shared_ptr**; §6.3, §15.2.1.
[10] use **unique_ptr** or **shared_ptr** as arguments or return values only to transfer ownership responsibilities; §15.2.1; [CG: F.26] [CG: F.27].
[11] Use **make_unique()** to construct **unique_ptr**s; §15.2.1; [CG: R.22].
[12] Use **make_shared()** to construct **shared_ptr**s; §15.2.1; [CG: R.23].
[13] Prefer smart pointers to garbage collection; §6.3, §15.2.1.
[14] Prefer **span**s to pointer-plus-count interfaces; §15.2.2; [CG: F.24].
[15] **span** supports **range**-for; §15.2.2.
[16] Use **array** where you need a sequence with a **constexpr** size; §15.3.1.
[17] Prefer **array** over built-in arrays; §15.3.1; [CG: SL.con.2].
[18] Use **bitset** if you need **N** bits and **N** is not necessarily the number of bits in a built-in integer type; §15.3.2.
[19] Don't overuse **pair** and **tuple**; named **struct**s often lead to more readable code; §15.3.3.
[20] When using **pair**, use template argument deduction or **make_pair()** to avoid redundant type specification; §15.3.3.
[21] When using **tuple**, use template argument deduction or **make_tuple()** to avoid redundant type specification; §15.3.3; [CG: T.44].
[22] Prefer **variant** to explicit use of **union**s; §15.4.1; [CG: C.181].
[23] When selecting among a set of alternatives using a **variant**, consider using **visit()** and **overloaded()**; §15.4.1.
[24] If more than one alternative is possible for a **variant**, **optional**, or **any**, check the tag before access; §15.4.

16

Utilities
（实用工具）

The time you enjoy wasting
is not wasted time.
– Bertrand Russell

- Introduction
- Time
  Clocks; Calendars; Time Zones
- Function Adaption
  Lambdas as Adaptors; **mem_fn()**; **function**
- Type Functions
  Type Predicates; Conditional Properties; Type Generators; Associate Types
- **source_location**
- **move()** and **forward()**
- Bit Manipulation
- Exiting a Program
- Advice

16.1 Introduction（引言）

Labeling a library component as a "Utility" isn't very informative. Obviously, every library component has been of utility to someone, somewhere, at some point in time. The facilities presented here are chosen because they serve critical purposes for many but their description doesn't fit elsewhere. Often, they act as building blocks for more powerful library facilities, including other components of the standard library.

16.2 Time（时间）

In <chrono>, the standard library provides facilities for dealing with time:

- Clocks, **time_point**, and **duration** for measuring how long some action takes, and as the basis for anything to do with time.
- **day**, **month**, **year**, and **weekdays** for mapping **time_points** into our everyday lives.
- **time_zone** and **zoned_time** to deal with differences in time reporting across the globe.

Essentially every major system deals with some of these entities.

16.2.1 Clocks（时钟）

Here is the basic way of timing some action:

```
using namespace std::chrono;          // in sub-namespace std::chrono; see §3.3

auto t0 = system_clock::now();
do_work();
auto t1 = system_clock::now();

cout << t1-t0 << "\n";                                              // default unit: 20223[1/00000000]s
cout << duration_cast<milliseconds>(t1-t0).count() << "ms\n";      // specify unit: 2ms
cout << duration_cast<nanoseconds>(t1-t0).count() << "ns\n";       // specify unit: 2022300ns
```

The clock returns a **time_point** (a point in time). Subtracting two **time_points** gives a **duration** (a period of time). The default << for a **duration** adds some indication of the unit used as a suffix. Various clocks give their results in various units of time, "clock ticks," (the clock I used measures in hundreds of nanoseconds), so it is often a good idea to convert a **duration** into an appropriate unit. That's what **duration_cast** does.

The clocks are useful for quick measurements. Don't make statements about "efficiency" of code without first doing time measurements. Guesses about performance are most unreliable. Quick, simple measurements are better than no measurements, but performance of modern computers is a tricky topic, so we must be careful not to attach too much importance to a few simple measurements. Always measure repeatedly to lower the chance of getting blindsided by rare events or cache effects.

Namespace **std::chrono_literals** defines time-unit suffixes (§6.6). For example:

```
this_thread::sleep_for(10ms+33us);     // wait for 10 milliseconds and 33 microseconds
```

Conventional symbolic names greatly increases readability and makes code more maintainable.

16.2.2 Calendars（日历）

When dealing with everyday events, we rarely use milliseconds; we use years, months, days, hours, seconds, and days of the week. The standard library supports that. For example:

```
auto spring_day = April/7/2018;
cout << weekday(spring_day) << '\n';                    // Sat
cout << format("{:%A}\n",weekday(spring_day));          // Saturday
```

Sat is the default character representation of Saturday on my computer. I didn't like that abbreviation, so I used **format** (§11.6.2) to get the longer name. For obscure reasons, **%A** means "write the day of the week's full name." Naturally, **April** is a month; more precisely a **std::chrono::Month**. We could also say

```
auto spring_day = 2018y/April/7;
```

The **y** suffix is used to distinguish years from plain **int**s which are used for days of the month numbered 1 to 31.

It is possible to express invalid dates. If in doubt, check with **ok()**:

```
auto bad_day = January/0/2024;
if (!bad_day.ok())
      cout << bad_day << " is not a valid day\n";
```

Obviously, **ok()** is most useful for dates obtained from a computation.

Dates are composed by overloading operator **/** (slash) by the types **year**, **month**, and **int**. The resulting **Year_month_day** type has conversions to and from **time_point** to allow accurate and efficient computation involving dates. For example:

```
sys_days t = sys_days{February/25/2022};    // get a time point with the precision of days
t += days{7};                               // one week after February 25, 2022
auto d = year_month_day(t);                 // convert the time point back to the calendar

cout << d << '\n';                                            // 2022-03-04
cout << format("{:%B}/{}/{}\n", d.month(), d.day(), d.year()); // March/04/2022
```

This calculation requires a change of month and knowledge about leap years. By default, the implementation gave the date in ISO 8601 standard format. To get the month spelled out as "March," we have to break out the individual fields of the date and get into formatting details (§11.6.2). For obscure reasons, **%B** means "write the month's full name."

Such operations can often be done at compile time and are therefore surprisingly fast:

```
static_assert(weekday(April/7/2018) == Saturday); // true
```

Calendars are complex and subtle. That's typical of and appropriate for "systems" designed for "ordinary people" over centuries, rather than by programmers to simplify programming. The standard-library calendar system can be (and has been) extended to cope with Julian, Islamic, Thai, and other calendars.

16.2.3 Time Zones（时区）

One of the trickiest issues to get right in connection with time is time zones. They are so arbitrary that they are hard to remember and they change in a variety of ways at times that are not standardized across the globe. For example:

```
auto tp = system_clock::now();                    // tp is a time_point
cout << tp << '\n';                               // 2021-11-27 21:36:08.2085095

zoned_time ztp { current_zone(),tp };             // 2021-11-27 16:36:08.2085095 EST
cout << ztp << '\n';

const time_zone est {"Europe/Copenhagen"};
cout << zoned_time{ &est,tp } << '\n';            // 2021-11-27 22:36:08.2085095 GMT+1
```

A **time_zone** is a time relative to a standard (called GMT or UTC) used by the **system_clock**. The standard library synchronizes with a global data base (IANA) to get its answers right. That synchronization can be automatic in the operating system or under the control of a system administrator. The names of time zones are C-style strings of the form "continent / major city", such as **"America/New_York"**, **"Asia/Tokyo"**, **"Africa/Nairobi"**. A **zoned_time** is a **time_zone** together with a **time_point**.

Like calendars, time zones address a set of concerns that we should leave to the standard library, rather than rely on our own handcrafted code. Consider: At what time of day on the last day of February 2024 in New York will the date change in New Delhi? When did summer-time (daylight savings time) end in Denver, Colorado, USA in 2020? When will the next leap second occur? The standard library "knows."

16.3 Function Adaption（函数适配）

When passing a function as a function argument, the type of the argument must exactly match the expectations expressed in the called function's declaration. If an intended argument only "almost matches expectations," we have alternative ways of adjusting it:

- Use a lambda (§16.3.1).
- Use **std::mem_fn()** to make a function object from a member function (§16.3.2).
- Define the function to accept a **std::function** (§16.3.3).

There are many other ways, but usually one of these three ways works best.

16.3.1 Lambdas as Adaptors（匿名函数作为适配器）

Consider the classical "draw all shapes" example:

```
void draw_all(vector<Shape*>& v)
{
    for_each(v.begin(),v.end(),[](Shape* p) { p->draw(); });
}
```

Like all standard-library algorithms, **for_each()** calls its argument using the traditional function call syntax **f(x)**, but **Shape's draw()** uses the conventional object-oriented notation **x->f()**. A lambda easily mediates between the two notations.

16.3.2 mem_fn()

Given a member function, the function adaptor **mem_fn(mf)** produces a function object that can be called as a nonmember function. For example:

```
void draw_all(vector<Shape*>& v)
{
     for_each(v.begin(),v.end(),mem_fn(&Shape::draw));
}
```

Before the introduction of lambdas in C++11, **mem_fn()** and equivalents were the main way to map from the object-oriented calling style to the functional one.

16.3.3 function

The standard-library **function** is a type that can hold any object you can invoke using the call operator **()**. That is, an object of type **function** is a function object (§7.3.2). For example:

```
int f1(double);
function<int(double)> fct1 {f1};                // initialize to f1

int f2(string);
function fct2 {f2};                             // fct2's type is function<int(string)>

function fct3 = [](Shape* p) { p->draw(); };    // fct3's type is function<void(Shape*)>
```

For **fct2**, I let the type of the **function** be deduced from the initializer: **int(string)**.

Obviously, **functions** are useful for callbacks, for passing operations as arguments, for passing function objects, etc. However, it may introduce some run-time overhead compared to direct calls. In particular, for a **function** object for which the size is not computed at compile-time, a free-store allocation might occur with seriously bad implications to performance-critical applications. A solution is coming for C++23: **move_only_function**.

Another problem is that **function**, being an object, does not participate in overloading. If you need to overload function objects (including lambdas), consider **overloaded** (§15.4.1).

16.4 Type Functions（类型函数）

A *type function* is a function that is evaluated at compile time given a type as its argument or returning a type. The standard library provides a variety of type functions to help library implementers (and programmers in general) write code that takes advantage of aspects of the language, the standard library, and code in general.

For numerical types, **numeric_limits** from **<limits>** presents a variety of useful information (§17.7). For example:

```
constexpr float min = numeric_limits<float>::min();      // smallest positive float
```

Similarly, object sizes can be found by the built-in **sizeof** operator (§1.4). For example:

```
constexpr int szi = sizeof(int);        // the number of bytes in an int
```

In <type_traits>, the standard library provides many functions for inquiring about properties of types. For example:

```
bool b = is_arithmetic_v<X>;                  // true if X is one of the (built-in) arithmetic types
using Res = invoke_result_t<decltype(f)>;     // Res is int if f is a function that returns an int
```

The **decltype(f)** is a call of the built-in type function **decltype()** returning the declared type of its argument; here **f**.

Some type functions create new types based on inputs. For example:

```
typename<typename T>
using Store = conditional_t(sizeof(T)<max, On_stack<T>, On_heap<T>);
```

If the first (Boolean) argument to **conditional_t** is **true**, the result is the first alternative; otherwise, the second. Assuming that **On_stack** and **On_heap** offer the same access functions to **T**, they can allocate their **T** as their names indicate. Thus, users of **Store<X>** can be tuned according to the size of **X** objects. The performance tuning enabled by this choice of allocation can be very significant. This is a simple example of how we can construct our own type functions, either from the standard ones or by using concepts.

Concepts are type functions. When used in expressions, they are specifically type predicates. For example:

```
template<typename F, typename... Args>
auto call(F f, Args... a, Allocator alloc)
{
    if constexpr (invocable<F,alloc,Args...>) // needs an allocator?
        return f(f,alloc,a...);
    else
        return f(f,a...);
}
```

In many cases, concepts are the best type functions, but most of the standard library was written pre-concepts and must support pre-concept code bases.

The notational conventions are confusing. The standard library uses **_v** for type functions that return values, _t for type functions that return types, This is a leftover from the weakly typed days of C and pre-concept C++. No standard-library type function returns both a type and a value, so these suffixes are redundant. With concepts, both in the standard library and elsewhere, no suffix is needed or used.

Type functions are part of C++'s mechanisms for compile-time computation that allow tighter type checking and better performance than would have been possible without them. Use of type functions and concepts (Chapter 8, §14.5) is often called *metaprogramming* or (when templates are involved) *template metaprogramming*.

16.4.1 Type Predicates（类型谓词）

In <type_traits>, the standard library offers dozens of simple type functions, called *type predicates* that answer fundamental questions about types. Here is a small selection:

Selected Type Predicates T, A, and U are types; all predicates returns a bool	
is_void_v<T>	is T void?
is_integral_v<T>	is T an integral type?
is_floating_point_v<T>	is T a floating-point type?
is_class_v<T>	is T a class (and not a union)?
is_function_v<T>	is T a function (and not a function object or a pointer to function)?
is_arithmetic_v<T>	is T an integral or floating-point type?
is_scalar_v<T>	is T an arithmetic, enumeration, pointer, or pointer to member type?
is_constructible_v<T, A...>	can a T be constructed from the A... argument list?
is_default_constructible_v<T>	can a T be constructed without explicit arguments?
is_copy_constructible_v<T>	can a T be constructed from another T?
is_move_constructible_v<T>	can a T be moved or copied into another T?
is_assignable_v<T,U>	can a U be assigned to a T?
is_trivially_copyable_v<T,U>	can a U be assigned to a T without user-defined copy operations?
is_same_v<T,U>	is T the same type as U?
is_base_of_v<T,U>	is U derived from T or is U the same type as U?
is_convertible_v<T,U>	can a T be implicitly converted to a U?
is_iterator_v<T>	is T an iterator type?
is_invocable_v<T, A...>	can a T be called with the argument list A...?
has_virtual_destructor_v<T>	does T have a virtual destructor?

One traditional use of these predicates is to constrain template arguments. For example:

```
template<typename Scalar>
class complex {
     Scalar re, im;
public:
     static_assert(is_arithmetic_v<Scalar>, "Sorry, I support only complex of arithmetic types");
     // ...
};
```

However, that – like other traditional uses – is easier and more elegantly done using concepts:

```
template<Arithmetic Scalar>
class complex {
     Scalar re, im;
public:
     // ...
};
```

In many cases, type predicates such as is_arithmetic disappear into the definition of concepts for greater ease of use. For example:

```
template<typename T>
concept Arithmetic = is_arithmetic_v<T>;
```

Curiously enough, there is no std::arithmetic concept.

Often, we can define concepts that are more general than the standard-library type predicates. Many standard-library type predicates apply only to the built-in types. We can define a concept in terms of the operations required, as suggested by the definition of **Number** (§8.2.4):

```
template<typename T, typename U = T>
concept Arithmetic = Number<T,U> && Number<U,T>;
```

Most often, uses of the standard-library type predicates are found deep in the implementation of fundamental services, often to distinguish cases for optimization. For example, part of the implementation of **std::copy(Iter,Iter,Iter2)** could optimize the important case of contiguous sequences of simple types, such as integers:

```
template<class T>
void cpy1(T* first, T* last, T* target)
{
	if constexpr (is_trivially_copyable_v<T>)
		memcpy(first, target, (last - first) * sizeof(T));
	else
		while (first != last) *target++ = *first++;
}
```

That simple optimization beat its non-optimized variant by about 50% on some implementations. Do not indulge in such cleverness unless you have verified that the standard doesn't already do better. Hand-optimized code is typically less maintainable than simpler alternatives.

16.4.2 Conditional Properties（条件属性）

Consider defining a "smart pointer":

```
template<typename T>
class Smart_pointer {
	// ...
	T& operator*() const;
	T* operator->() const;	// -> should work if and only if T is a class
};
```

The -> should be defined if and only if **T** is a class type. For example, **Smart_pointer<vector<T>>** should have ->, but **Smart_pointer<int>** should not.

We cannot use a compile-time **if** because we are not inside a function. Instead, we write

```
template<typename T>
class Smart_pointer {
	// ...
	T& operator*() const;
	T* operator->() const requires is_class_v<T>;		// -> is defined if and only if T is a class
};
```

The type predicate directly expresses the constraint on **operator->()**. We can also use concepts for that. There is no standard-library concept for requiring a type to be a class type (that is, a **class**, a **struct**, or **union**), but we could define one:

```
template<typename T>
concept Class = is_class_v<T> || is_union_v<T>;          // unions are classes

template<typename T>
class Smart_pointer {
     // ...
     T& operator*() const;
     T* operator->() const requires Class<T>;           // -> is defined if and only if T is a class or a union
};
```

Often, a concept is more general or simply more appropriate than the direct use of a standard-library type predicate.

16.4.3 Type Generators（类型生成器）

Many type functions return types, often new types that they compute. I call such functions *type generators* to distinguish them from the type predicates. The standard offers a few, such as:

Selected Type Generators	
R=remove_const_t<T>	R is T with the topmost const (if any) removed
R=add_const_t<T>	R is const T
R=remove_reference_t<T>	if T is a reference U&, R is U otherwise T
R=add_lvalue_reference_t<T>	if T is an lvalue reference, R is T otherwise T&
R=add_rvalue_reference_t<T>	if T is an rvalue reference, R is T otherwise T&&
R=enable_if_t<b,T =void>	if b is true, R is T otherwise R is not defined
R=conditional_t<b,T,U>	R is T if b is true; U otherwise
R=common_type_t<T...>	if there is a type that all Ts can be implicitly converted to, R is that type; otherwise R is not defined
R=underlying_type_t<T>	if T is an enumeration, R is its underlying type; otherwise error
R=invoke_result_t<T,A...>	if a T can be called with arguments A..., R is its return type; otherwise error

These type functions are typically used in the implementation of utilities, rather than directly in application code. Of these, enable_if is probably the most common in pre-concepts code. For example, the conditionally enabled -> for a smart pointer is traditionally implemented something like this:

```
template<typename T>
class Smart_pointer {
     // ...
     T& operator*();
     enable_if<is_class_v<T>,T&> operator->();     // -> is defined if and only if T is a class
};
```

I don't find this particularly easy to read and more complicated uses are far worse. The definition of enable_if relies on a subtle language feature called SFINAE ("Substitution Failure Is Not An Error"). Look that up (only) if you need to.

16.4.4 Associated Types（关联类型）

All standard containers (§12.8) and all containers designed to follow their pattern have some *associated types* such as their value types and iterator types. In <iterator> and <ranges>, the standard library supplies names for those:

	Selected Type Generators
range_value_t<R>	The type of the range R's elements
iter_value_t<T>	The type of elements pointed to by the iterator T
iterator_t<R>	The type of the range R's iterator

16.5 source_location

When writing out a trace message or an error message, we often want to make a source location part of that message. The library provides source_location for that:

```
const source_location loc = source_location::current();
```

That current() returns a source_location describing the spot in the source code where it appears. Class source_location has file() and function_name() members returning C-style strings and line() and column() members returning unsigned integers.

Wrap that in a function and we have a good first cut of a logging message:

```
void log(const string& mess = "", const source_location loc = source_location::current())
{
	cout << loc.file_name()
		<< '(' << loc.line() << ':' << loc.column() << ") "
		<< loc.function_name() ": "
		<< mess;
}
```

The call of current() is a default argument so that we get the location of the caller of log() rather than the location of log():

```
void foo()
{
	log("Hello");		// myfile.cpp (17,4) foo: Hello
	// ...
}

int bar(const string& label)
{
	log(label);		// myfile.cpp (23,4) bar: <<the value of label>>
	// ...
}
```

Code written before C++20 or needing to compile on older compilers uses macros __FILE__ and __LINE__ for this.

16.6 move() and forward()（move()和 forward()）

The choice between moving and copying is mostly implicit (§3.4). A compiler will prefer to move when an object is about to be destroyed (as in a **return**) because that's assumed to be the simpler and more efficient operation. However, sometimes we must be explicit. For example, a **unique_ptr** is the sole owner of an object. Consequently, it cannot be copied, so if you want a **unique_ptr** elsewhere, you must move it. For example:

```
void f1()
{
      auto p = make_unique<int>(2);
      auto q = p;                       // error: we can't copy a unique_ptr
      auto q = move(p);                 // p now holds nullptr
      // ...
}
```

Confusingly, **std::move()** doesn't move anything. Instead, it casts its argument to an rvalue reference, thereby saying that its argument will not be used again and therefore may be moved (§6.2.2). It should have been called something like **rvalue_cast**. It exists to serve a few essential cases. Consider a simple swap:

```
template <typename T>
void swap(T& a, T& b)
{
      T tmp {move(a)};                  // the T constructor sees an rvalue and moves
      a = move(b);                      // the T assignment sees an rvalue and moves
      b = move(tmp);                    // the T assignment sees an rvalue and moves
}
```

We don't want to repeatedly copy potentially large objects, so we request moves using **std::move()**.

As for other casts, there are tempting, but dangerous, uses of **std::move()**. Consider:

```
string s1 = "Hello";
string s2 = "World";
vector<string> v;
v.push_back(s1);                 // use a "const string&" argument; push_back() will copy
v.push_back(move(s2));           // use a move constructor
v.emplace_back(s1);              // an alternative; place a copy of s1 in a new end position of v (§12.8)
```

Here **s1** is copied (by **push_back()**) whereas **s2** is moved. This sometimes (only sometimes) makes the **push_back()** of **s2** cheaper. The problem is that a moved-from object is left behind. If we use **s2** again, we have a problem:

```
cout << s1[2];          // write 'l'
cout << s2[2];          // crash?
```

I consider this use of **std::move()** to be too error-prone for widespread use. Don't use it unless you can demonstrate significant and necessary performance improvement. Later maintenance may accidentally lead to unanticipated use of the moved-from object.

The compiler knows that a return value is not used again in a function, so using an explicit **std::move()**, e.g., **return std::move(x)**, is redundant and can even inhibit optimizations.

The state of a moved-from object is in general unspecified, but all standard-library types leave a moved-from object in a state where it can be destroyed and assigned to. It would be unwise not to follow that lead. For a container (e.g., **vector** or **string**), the moved-from state will be "empty." For many types, the default value is a good empty state: meaningful and cheap to establish.

Forwarding arguments is an important use case that requires moves (§8.4.2). We sometimes want to transmit a set of arguments on to another function without changing anything (to achieve "perfect forwarding"):

```
template<typename T, typename... Args>
unique_ptr<T> make_unique(Args&&... args)
{
     return unique_ptr<T>{new T{std::forward<Args>(args)...}};     // forward each argument
}
```

The standard-library **forward()** differs from the simpler **std::move()** by correctly handling subtleties to do with lvalue and rvalue (§6.2.2). Use **std::forward()** exclusively for forwarding and don't **forward()** something twice; once you have forwarded an object, it's not yours to use anymore.

16.7 Bit Manipulation（位操作）

In **<bit>**, we find functions for low-level bit manipulation. Bit manipulation is a specialized, but often essential, activity. When we get close to the hardware, we often have to look at bits, change bit patterns in a byte or a word, and turn raw memory into typed objects. For example, **bit_cast** lets us convert a value of one type to another type of the same size:

```
double val = 7.2;
auto x = bit_cast<uint64_t>(val);     // get the bit representation of a 64-bit floating point number
auto y = bit_cast<uint64_t>(&val);    // get the bit representation of a 64-bit pointer

struct Word { std::byte b[8]; };
std::byte buffer[1024];
// ...
auto p = bit_cast<Word*>(&buffer[i]);     // p points to 8 bytes
auto i = bit_cast<int64_t>(*p);            // convert those 8 bytes to an integer
```

The standard-library type **std::byte** (the **std::** is required) exists to represent bytes, rather than bytes known to represent characters or integers. In particular, **std::byte** provides only bit-wise logical operations and not arithmetic ones. Usually, the best type to do bit operations on is an unsigned integer or **std::byte**. By best, I mean fastest and least likely to surprise. For example:

```
void use(unsigned int ui)
{
     int x0 = bit_width(ui)                // the smallest number of bits needed to represent ui
     unsigned int ui2 = rotl(ui,8)         // rotate left 8 bits (note: doesn't change ui)
     int x1 = popcount(ui);                // the number of 1s in ui
     // ...
}
```

See also **bitset** (§15.3.2).

16.8 Exiting a Program（退出程序）

Occasionally, a piece of code encounters a problem that it cannot handle:

- If the kind of problem is frequent and the immediate caller can be expected to handle it, return some kind of return code (§4.4).
- If the kind of problem is infrequent or the immediate caller cannot be expected to handle it, throw an exception (§4.4).
- If the kind of problem is so serious that no ordinary part of the program can be expected to handle it, exit the program.

The standard library provides facilities to deal with that last case ("exit the program"):

- **exit(x)**: call functions registered with **atexit()** then exit the program with the return value **x**. If you need to, look up **atexit()**, it's basically a primitive destructor mechanism shared with the C language.
- **abort()**: exit the program immediately and unconditionally with a return value indicating unsuccessful termination. Some operating systems offer facilities that modify this simple explanation.
- **quick_exit(x)**: call functions registered with **at_quick_exit()**; then exit the program with the return value **x**.
- **terminate()**: call the **terminate_handler**. The default **terminate_handler** is **abort()**.

These functions are for really serious errors. They do not invoke destructors; that is, they do not do ordinary and proper clean-up. The various handlers are used to take actions before exiting. Such actions must be very simple because one reason for calling these exit functions is that the program state is corrupted. One reasonable and reasonably popular action is "restart the system in a well-defined state relying on no state from the current program." Another, slightly dicier, but often not unreasonable action is "log an error message and exit." The reason that writing a logging message can be a problem is that the I/O system might have been corrupted by whatever caused the exit function to be called.

Error handling is one of the trickiest kinds of programming; even getting cleanly out of a program can be hard.

No general-purpose library should unconditionally terminate.

16.9 Advice（建议）

[1] A library doesn't have to be large or complicated to be useful; §16.1.
[2] Time your programs before making claims about efficiency; §16.2.1.
[3] Use **duration_cast** to report time measurements with proper units; §16.2.1.
[4] To represent a date directly in source code, use symbolic notation (e.g., **November/28/2021**); §16.2.2.
[5] If a date is a result of a computation, check for validity using **ok()**; §16.2.2.
[6] When dealing with time in different locations, use **zoned_time**; §16.2.3.
[7] Use a lambda to express minor changes in calling conventions; §16.3.1.
[8] Use **mem_fn()** or a lambda to create function objects that can invoke a member function when called using the traditional function call notation; §16.3.1, §16.3.2.

[9] Use **function** when you need to store something that can be called; §16.3.3.
[10] Prefer concepts to explicit use of type predicates; §16.4.1.
[11] You can write code to explicitly depend on properties of types; §16.4.1, §16.4.2.
[12] Prefer concepts over traits and **enable_if** whenever you can; §16.4.3.
[13] Use **source_location** to embed source code locations in debug and logging messages; §16.5.
[14] Avoid explicit use of **std::move()**; §16.6; [CG: ES.56].
[15] Use **std::forward()** exclusively for forwarding; §16.6.
[16] Never read from an object after **std::move()**ing or **std::forward()**ing it; §16.6.
[17] Use **std::byte** to represent data that doesn't (yet) have a meaningful type; §16.7.
[18] Use **unsigned** integers or **bitset**s for bit manipulation §16.7.
[19] Return an error-code from a function if the immediate caller can be expected to handle the problem; §16.8.
[20] Throw an exception from a function if the immediate caller cannot be expected to handle the problem; §16.8.
[21] Call **exit()**, **quick_exit()**, or **terminate()** to exit a program if an attempt to recover from a problem is not reasonable; §16.8.
[22] No general-purpose library should unconditionally terminate; §16.8.

17

Numerics
（数值计算）

The purpose of computing is insight, not numbers.
– R. W. Hamming

... but for the student,
numbers are often the best road to insight.
– A. Ralston

17.1 Introduction（引言）

C++ was not designed primarily with numeric computation in mind. However, numeric computation typically occurs in the context of other work – such as scientific computation, database access, networking, instrument control, graphics, simulation, and financial analysis – so C++ becomes an attractive vehicle for computations that are part of a larger system. Furthermore, numeric methods have come a long way from being simple loops over vectors of floating-point numbers. Where more complex data structures are needed as part of a computation, C++'s strengths become relevant. The net effect is that C++ is widely used for scientific, engineering, financial, and other computation involving sophisticated numerics. Consequently, facilities and techniques supporting such computation have emerged. This chapter describes the parts of the standard library that support numerics.

17.2 Mathematical Functions（数学函数）

In <cmath>, we find the *standard mathematical functions*, such as **sqrt()**, **log()**, and **sin()** for arguments of type **float**, **double**, and **long double**:

Selected Standard Mathematical Functions	
abs(x)	Absolute value
ceil(x)	Smallest integer >= x
floor(x)	Largest integer <= x
sqrt(x)	Square root; x must be non-negative
cos(x)	Cosine
sin(x)	Sine
tan(x)	Tangent
acos(x)	Arccosine; the result is non-negative
asin(x)	Arcsine; the result nearest to 0 is returned
atan(x)	Arctangent
sinh(x)	Hyperbolic sine
cosh(x)	Hyperbolic cosine
tanh(x)	Hyperbolic tangent
exp(x)	Base e exponential
exp2(x)	Base 2 exponential
log(x)	Natural logarithm, base e; x must be positive
log2(x)	Natural logarithm, base 2; x must be positive
log10(x)	Base 10 logarithm; x must be positive

The versions of these functions for **complex** (§17.4) are found in <complex>. For each function, the return type is the same as the argument type.

Errors are reported by setting **errno** from <cerrno> to **EDOM** for a domain error and to **ERANGE** for a range error. For example:

```
errno = 0;        // clear old error state
double d = sqrt(-1);
if (errno==EDOM)
      cerr << "sqrt() not defined for negative argument\n";

errno = 0;        // clear old error state
double dd = pow(numeric_limits<double>::max(),2);
if (errno == ERANGE)
      cerr << "result of pow() too large to represent as a double\n";
```

More mathematical functions are found in <cmath> and <cstdlib>. The so-called *special mathematical functions*, such as **beta()**, **rieman_zeta()**, and **sph_bessel()**, are also in <cmath>.

17.3 Numerical Algorithms（数值计算算法）

In <numeric>, we find a small set of generalized numerical algorithms, such as accumulate().

Numerical Algorithms	
x=accumulate(b,e,i)	x is the sum of i and the elements of [b:e)
x=accumulate(b,e,i,f)	accumulate using f instead of +
x=inner_product(b,e,b2,i)	x is the inner product of [b:e) and [b2:b2+(e−b)), that is, the sum of i and (*p1)*(*p2) for each p1 in [b:e) and the corresponding p2 in [b2:b2+(e−b))
x=inner_product(b,e,b2,i,f,f2)	inner_product using f and f2 instead of + and *
p=partial_sum(b,e,out)	Element i of [out:p) is the sum of elements [b:b+i]
p=partial_sum(b,e,out,f)	partial_sum using f instead of +
p=adjacent_difference(b,e,out)	Element i of [out:p) is *(b+i)−*(b+i−1) for i>0; if e−b>0, then *out is *b
p=adjacent_difference(b,e,out,f)	adjacent_difference using f instead of –
iota(b,e,v)	For each element in [b:e) assign v and increment ++v; thus the sequence becomes v, v+1, v+2, ...
x=gcd(n,m)	x is the greatest common denominator of integers n and m
x=lcm(n,m)	x is the least common multiple of integers n and m
x=midpoint(n,m)	x is the midpoint between n and m

These algorithms generalize common operations such as computing a sum by letting them apply to all kinds of sequences. They also make the operation applied to elements of those sequences a parameter. For each algorithm, the general version is supplemented by a version applying the most common operator for that algorithm. For example:

```
list<double> lst {1, 2, 3, 4, 5, 9999.99999};
auto s = accumulate(lst.begin(),lst.end(),0.0);        // calculate the sum: 10014.9999
```

These algorithms work for every standard-library sequence and can have operations supplied as arguments (§17.3).

17.3.1 Parallel Numerical Algorithms（并行数值算法）

In <numeric>, the numerical algorithms (§17.3) have parallel versions that differ slightly from the sequentional ones. In particular, the parallel versions allow operations on elements in unspecified order. The parallel numerical algorithms can take an execution policy argument (§13.6): seq, unseq, par, and par_unseq.

Parallel Numerical Algorithms	
x=reduce(b,e,v)	x=accumulate(b,e,v), except out of order
x=reduce(b,e)	x=reduce(b,e,V{}), where V is b's value type
x=reduce(pol,b,e,v)	x=reduce(b,e,v) with execution policy pol
x=reduce(pol,b,e)	x=reduce(pol,b,e,V{}), where V is b's value type
p=exclusive_scan(pol,b,e,out)	p=partial_sum(b,e,out) according to pol, excludes the ith input element from the ith sum
p=inclusive_scan(pol,b,e,out)	p=partial_sum(b,e,out) according to pol includes the ith input element in the ith sum
p=transform_reduce(pol,b,e,f,v)	f(x) for each x in [b:e), then reduce
p=transform_exclusive_scan(pol,b,e,out,f,v)	f(x) for each x in [b:e), then exclusive_scan
p=transform_inclusive_scan(pol,b,e,out,f,v)	f(x) for each x in [b:e), then inclusive_scan

For simplicity, I left out the versions of these algorithms that take operations as arguments, rather than just using + and =. Except for reduce(), I also left out the versions with default policy (sequential) and default value.

Just as for the parallel algorithms in <algorithm> (§13.6), we can specify an execution policy:

```
vector<double> v {1, 2, 3, 4, 5, 9999.99999};
auto s = reduce(v.begin(),v.end());          // calculate the sum using a double as the accumulator

vector<double> large;
// ... fill large with lots of values ...
auto s2 = reduce(par_unseq,large.begin(),large.end()); // calculate the sum using available parallelism
```

The execution policies, par, sec, unsec, and par_unsec are hidden in namespace std::execution in <execution>.

Measure to verify that using a parallel or vectorized algorithm is worthwhile.

17.4 Complex Numbers（复数）

The standard library supports a family of complex number types along the lines of the complex class described in §5.2.1. To support complex numbers where the scalars are single-precision floating-point numbers (floats), double-precision floating-point numbers (doubles), etc., the standard library complex is a template:

```
template<typename Scalar>
class complex {
public:
     complex(const Scalar& re ={}, const Scalar& im ={});    // default function arguments; see §3.4.1
     // ...
};
```

The usual arithmetic operations and the most common mathematical functions are supported for complex numbers. For example:

```
void f(complex<float> fl, complex<double> db)
{
	complex<long double> ld {fl+sqrt(db)};
	db += fl*3;
	fl = pow(1/fl,2);
	// ...
}
```

The sqrt() and pow() (exponentiation) functions are among the usual mathematical functions defined in <complex> (§17.2).

17.5 Random Numbers（随机数）

Random numbers are useful in many contexts, such as testing, games, simulation, and security. The diversity of application areas is reflected in the wide selection of random number generators provided by the standard library in <random>. A random number generator consists of two parts:

[1] An *engine* that produces a sequence of random or pseudo-random values

[2] A *distribution* that maps those values into a mathematical distribution in a range

Examples of distributions are uniform_int_distribution (where all integers produced are equally likely), normal_distribution ("the bell curve"), and exponential_distribution (exponential growth); each for some specified range. For example:

```
using my_engine = default_random_engine;                 // type of engine
using my_distribution = uniform_int_distribution<>;      // type of distribution

my_engine eng {};                                        // the default version of the engine
my_distribution dist {1,6};                              // distribution that maps to the ints 1..6
auto die = [&](){ return dist(eng); };                   // make a generator

int x = die();                                           // roll the die: x becomes a value in [1:6]
```

Thanks to its uncompromising attention to generality and performance, one expert has deemed the standard-library random number component "what every random number library wants to be when it grows up." However, it can hardly be deemed "novice friendly." The using statements and the lambda make what is being done a bit more obvious.

For novices (of any background) the fully general interface to the random number library can be a serious obstacle. A simple uniform random number generator is often sufficient to get started. For example:

```
Rand_int rnd {1,10};     // make a random number generator for [1:10]
int x = rnd();           // x is a number in [1:10]
```

So, how could we get that? We have to get something that, like die(), combines an engine with a distribution inside a class Rand_int:

```
class Rand_int {
public:
    Rand_int(int low, int high) :dist{low,high} { }
    int operator()() { return dist(re); }          // draw an int
    void seed(int s) { re.seed(s); }               // choose new random engine seed
private:
    default_random_engine re;
    uniform_int_distribution<> dist;
};
```

That definition is still "expert level," but the *use* of `Rand_int()` is manageable in the first week of a C++ course for novices. For example:

```
int main()
{
    constexpr int max = 9;
    Rand_int rnd {0,max};                          // make a uniform random number generator

    vector<int> histogram(max+1);                  // make a vector of appropriate size
    for (int i=0; i!=200; ++i)
        ++histogram[rnd()];                        // fill histogram with the frequencies of numbers [0:max]

    for (int i = 0; i!=histogram.size(); ++i) {    // write out a bar graph
        cout << i << '\t';
        for (int j=0; j!=histogram[i]; ++j) cout << '*';
        cout << '\n';
    }
}
```

The output is a (reassuringly boring) uniform distribution (with reasonable statistical variation):

```
0	*********************
1	****************
2	*******************
3	********************
4	****************
5	***********************
6	**************************
7	***********
8	**********************
9	*************************
```

There is no standard graphics library for C++, so I use "ASCII graphics." Obviously, there are lots of open source and commercial graphics and GUI libraries for C++, but in this book I restrict myself to ISO standard facilities.

To get a repeated or different sequence of values, we *seed* the engine; that is, we give its internal state a new value. For example:

```
Rand_int rnd {10,20};
for (int i = 0; i<10; ++i) cout << rnd() << ' ';        // 16 13 20 19 14 17 10 16 15 14
cout << '\n';
rnd.seed(999);
for (int i = 0; i<10; ++i) cout << rnd() << ' ';        // 11 17 14 19 20 13 20 14 16 19
cout << '\n';
rnd.seed(999);
for (int i = 0; i<10; ++i) cout << rnd() << ' ';        // 11 17 14 19 20 13 20 14 16 19
cout << '\n';
```

Repeated sequences are important for deterministic debugging. Seeding with different values is important when we don't want repetition. If you need genuine random numbers, rather than a generated pseudo-random sequence, look to see how **random_device** is implemented on your machine.

17.6 Vector Arithmetic（向量算术）

The **vector** described in §12.2 was designed to be a general mechanism for holding values, to be flexible, and to fit into the architecture of containers, iterators, and algorithms. However, it does not support mathematical vector operations. Adding such operations to **vector** would be easy, but its generality and flexibility preclude optimizations that are often considered essential for serious numerical work. Consequently, the standard library provides (in **<valarray>**) a **vector**-like template, called **valarray**, that is less general and more amenable to optimization for numerical computation:

```
template<typename T>
class valarray {
    // ...
};
```

The usual arithmetic operations and the most common mathematical functions are supported for **valarray**s. For example:

```
void f(valarray<double>& a1, valarray<double>& a2)
{
    valarray<double> a = a1*3.14+a2/a1;          // numeric array operators *, +, /, and =
    a2 += a1*3.14;
    a = abs(a);
    double d = a2[7];
    // ...
}
```

The operations are vector operations; that is, they are applied to each element of the vectors involved.

In addition to arithmetic operations, **valarray** offers stride access to help implement multidimensional computations.

17.7 Numeric Limits（数值界限）

In <limits>, the standard library provides classes that describe the properties of built-in types – such as the maximum exponent of a float or the number of bytes in an int. For example, we can assert that a char is signed:

```
static_assert(numeric_limits<char>::is_signed,"unsigned characters!");
static_assert(100000<numeric_limits<int>::max(),"small ints!");
```

The second assert (only) works because numeric_limits<int>::max() is a constexpr function (§1.6).

We can define numeric_limits for our own user-defined types.

17.8 Type Aliases（类型别名）

The size of fundamental types, such as int and long long are implementation defined; that is, they may be different on different implementations of C++. If we need to be specific about the size of our integers, we can use aliases defined in <stdint>, such as int32_t and uint_least64_t. The latter means an unsigned integer with at least 64 bits.

The curious _t suffix is a relict from the days of C when it was deemed important to have a name reflect that it named an alias.

Other common aliases, such as size_t (the type returned by the sizeof operator) and ptrdiff_t (the type of the result of subtracting one pointer from another) can be found in <stddef>.

17.9 Mathematical Constants（数学常数）

When doing mathematical computations, we need common mathematical constants such as e, pi, and log2e. The standard library offers those and more. They come in two forms: a template that allows us to specify exact type (e.g., pi_v<T>) and a short name for the most common use (e.g., pi meaning pi_v<double>). For example:

```
void area(float r)
{
        using namespace std::numbers;     // this is where the mathematical constants are kept

        double d = pi*r*r;
        float f = pi_v<float>*r*r;

        // ...
}
```

In this case, the difference is small (we would have to print with precision 16 or so to see it), but in real physics calculation such differences quickly become significant. Other areas where the precision of constants matters are graphics and AI, where smaller representations of values are increasingly important.

In <numbers>, we find e (Euler's number), log2e (log2 of e), log10e (log10 of e), pi, inv_pi (1/pi), inv_sqrtpi (1/sqrt(pi)), ln2, ln10, sqrt2 (sqrt(2)), sqrt3 (sqrt(3)), inv_sqrt3 (1/sqrt3), egamma (the Euler-Mascheroni constant), and phi (the golden ratio).

Naturally, we would like more mathematical constants and constants for different domains. That's easily done because such constants are variable templates with specializations for double (or whatever type is most useful for a domain):

```
template<typename T>
constexpr T tau_v = 2*pi_v<T>;

constexpr double tau = tau_v<double>;
```

17.10 Advice（建议）

[1] Numerical problems are often subtle. If you are not 100% certain about the mathematical aspects of a numerical problem, either take expert advice, experiment, or do both; §17.1.
[2] Don't try to do serious numeric computation using only the bare language; use libraries; §17.1.
[3] Consider accumulate(), inner_product(), partial_sum(), and adjacent_difference() before you write a loop to compute a value from a sequence; §17.3.
[4] For larger amounts of data, try the parallel and vectorized algorithms; §17.3.1.
[5] Use std::complex for complex arithmetic; §17.4.
[6] Bind an engine to a distribution to get a random number generator; §17.5.
[7] Be careful that your random numbers are sufficiently random for your intended use; §17.5.
[8] Don't use the C standard-library rand(); it isn't insufficiently random for real uses; §17.5.
[9] Use valarray for numeric computation when run-time efficiency is more important than flexibility with respect to operations and element types; §17.6.
[10] Properties of numeric types are accessible through numeric_limits; §17.7.
[11] Use numeric_limits to check that the numeric types are adequate for their use; §17.7.
[12] Use aliases for integer types if you want to be specific about their sizes; §17.8.

18

Concurrency

（并发）

Keep it simple:
as simple as possible,
but no simpler.
– A. Einstein

- Introduction
- Tasks and **threads**
  - Passing Arguments; Returning Results
- Sharing Data
  - **mutexes** and Locks; **atomics**
- Waiting for Events
- Communicating Tasks
  - **future** and **promise**; **packaged_task**; **async()**; Stopping a **thread**
- Coroutines
  - Cooperative Multitasking
- Advice

18.1 Introduction （引言）

Concurrency – the execution of several tasks simultaneously – is widely used to improve throughput (by using several processors for a single computation) or to improve responsiveness (by allowing one part of a program to progress while another is waiting for a response). All modern programming languages provide support for this. The support provided by the C++ standard library is a portable and type-safe variant of what has been used in C++ for more than 20 years and is almost universally supported by modern hardware. The standard-library support is primarily aimed at supporting systems-level concurrency rather than directly providing sophisticated higher-level concurrency models; those can be supplied as libraries built using the standard-library facilities.

The standard library directly supports concurrent execution of multiple threads in a single address space. To allow that, C++ provides a suitable memory model and a set of atomic operations. The atomic operations allow lock-free programming [Dechev,2010]. The memory model ensures that as long as a programmer avoids data races (uncontrolled concurrent access to mutable data), everything works as one would naively expect. However, most users will see concurrency only in terms of the standard library and libraries built on top of that. This section briefly gives examples of the main standard-library concurrency support facilities: **threads**, **mutexes**, **lock()** operations, **packaged_tasks**, and **futures**. These features are built directly upon what operating systems offer and do not incur performance penalties compared with those. Neither do they guarantee significant performance improvements compared to what the operating system offers.

Do not consider concurrency a panacea. If a task can be done sequentially, it is often simpler and faster to do so. Passing information from one thread to another can be surprisingly expensive.

As an alternative to using explicit concurrency features, we can often use a parallel algorithm to exploit multiple execution engines for better performance (§13.6, §17.3.1).

Finally, C++ supports coroutines; that is, functions that keep their state between calls (§18.6).

18.2 Tasks and threads（任务和 thread）

We call a computation that can potentially be executed concurrently with other computations a *task*. A *thread* is the system-level representation of a task in a program. A task to be executed concurrently with other tasks is launched by constructing a **thread** (found in **<thread>**) with the task as its argument. A task is a function or a function object:

```
void f();                    // function

struct F {                   // function object
      void operator()();     // F's call operator (§7.3.2)
};

void user()
{
      thread t1 {f};         // f() executes in separate thread
      thread t2 {F{}};       // F{}() executes in separate thread

      t1.join();             // wait for t1
      t2.join();             // wait for t2
}
```

The **join()**s ensure that we don't exit **user()** until the threads have completed. To "join" a **thread** means to "wait for the thread to terminate."

It is easy to forget to **join()**, and the results are usually bad so the standard library provides **jthread**, which is a "joining **thread**" that follows RAII by having its destructor **join()**:

```
void user()
{
    jthread t1 {f};          // f() executes in separate thread
    jthread t2 {F{}};            // F{}() executes in separate thread
}
```

Joining is done by destructors, so the order is the reverse of construction. Here, we wait for t2 before t1.

Threads of a program share a single address space. In this, threads differ from processes, which generally do not directly share data. Since threads share an address space, they can communicate through shared objects (§18.3). Such communication is typically controlled by locks or other mechanisms to prevent data races (uncontrolled concurrent access to a variable).

Programming concurrent tasks can be *very* tricky. Consider possible implementations of the tasks f (a function) and F (a function object):

```
void f()
{
    cout << "Hello ";
}

struct F {
    void operator()() { cout << "Parallel World!\n"; }
};
```

This is an example of a bad error: here, f and F{} each use the object cout without any form of synchronization. The resulting output would be unpredictable and could vary between different executions of the program because the order of execution of the individual operations in the two tasks is not defined. The program may produce "odd" output, such as

```
PaHerallllel o World!
```

Only a specific guarantee in the standard saves us from a data race within the definition of ostream that could lead to a crash.

To avoid such problems with output streams, either have just one thread use a stream or use a osyncstream (§11.7.5)

When defining tasks of a concurrent program, our aim is to keep tasks completely separate except where they communicate in simple and obvious ways. The simplest way of thinking of a concurrent task is as a function that happens to run concurrently with its caller. For that to work, we just have to pass arguments, get a result back, and make sure that there is no use of shared data in between (no data races).

18.2.1 Passing Arguments（传递参数）

Typically, a task needs data to work upon. We can easily pass data (or pointers or references to the data) as arguments. Consider:

```
void f(vector<double>& v);      // function: do something with v

struct F {                      // function object: do something with v
    vector<double>& v;
    F(vector<double>& vv) :v{vv} { }
    void operator()();          // application operator; §7.3.2
};

int main()
{
    vector<double> some_vec {1, 2, 3, 4, 5, 6, 7, 8, 9};
    vector<double> vec2 {10, 11, 12, 13, 14};

    jthread t1 {f,ref(some_vec)};   // f(some_vec) executes in a separate thread
    jthread t2 {F{vec2}};           // F(vec2)() executes in a separate thread
}
```

F{vec2} saves a reference to the argument vector in **F**. **F** can now use that vector and hopefully no other task accesses **vec2** while **F** is executing. Passing **vec2** by value would eliminate that risk.

The initialization with **{f,ref(some_vec)}** uses a **thread** variadic template constructor that can accept an arbitrary sequence of arguments (§8.4). The **ref()** is a type function from **<functional>** that unfortunately is needed to tell the variadic template to treat **some_vec** as a reference, rather than as an object. Without that **ref()**, **some_vec** would be passed by value. The compiler checks that the first argument can be invoked given the following arguments and builds the necessary function object to pass to the thread. Thus, if **F::operator()()** and **f()** perform the same algorithm, the handling of the two tasks is roughly equivalent: in both cases, a function object is constructed for the **thread** to execute.

18.2.2 Returning Results（返回结果）

In the example in §18.2.1, I pass the arguments by non-**const** reference. I only do that if I expect the task to modify the value of the data referred to (§1.7). That's a somewhat sneaky, but not uncommon, way of returning a result. A less obscure technique is to pass the input data by **const** reference and to pass the location of a place to deposit the result as a separate argument:

```
void f(const vector<double>& v, double* res);     // take input from v; place result in *res

class F {
public:
    F(const vector<double>& vv, double* p) :v{vv}, res{p} { }
    void operator()();              // place result in *res
private:
    const vector<double>& v;        // source of input
    double* res;                    // target for output
};

double g(const vector<double>&); // use return value
```

```
void user(vector<double>& vec1, vector<double> vec2, vector<double> vec3)
{
	double res1;
	double res2;
	double res3;

	thread t1 {f,cref(vec1),&res1};		// f(vec1,&res1) executes in a separate thread
	thread t2 {F{vec2,&res2}};		// F{vec2,&res2}() executes in a separate thread
	thread t3 { [&](){ res3 = g(vec3); } };	// capture local variables by reference

	t1.join();	// join before using results
	t2.join();
	t3.join();

	cout << res1 << ' ' << res2 << ' ' << res3 << '\n';
}
```

Here, **cref(vec1)** passes a **const** reference to **vec1** as an argument to **t1**.

This works and the technique is very common, but I don't consider returning results through references particularly elegant, so I return to this topic in §18.5.1.

18.3 Sharing Data（共享数据）

Sometimes tasks need to share data. In that case, the access has to be synchronized so that at most one task at a time has access. Experienced programmers will recognize this as a simplification (e.g., there is no problem with many tasks simultaneously reading immutable data), but consider how to ensure that at most one task at a time has access to a given set of objects.

18.3.1 mutexes and Locks（mutex 和锁）

A **mutex**, a "mutual exclusion object," is a key element of general sharing of data between **threads**. A **thread** acquires a **mutex** using a **lock()** operation:

```
mutex m;	// controlling mutex
int sh;		// shared data

void f()
{
	scoped_lock lck {m};	// acquire mutex
	sh += 7;		// manipulate shared data
}	// release mutex implicitly
```

The type of **lck** is deduced to be **scoped_lock<mutex>** (§7.2.3). The **scoped_lock**'s constructor acquires the mutex (through a call **m.lock()**). If another thread has already acquired the mutex, the thread waits ("blocks") until the other thread completes its access. Once a thread has completed its access to the shared data, the **scoped_lock** releases the **mutex** (with a call **m.unlock()**). When a **mutex** is released, **threads** waiting for it resume executing ("are woken up"). The mutual exclusion and locking facilities are found in **<mutex>**.

. Note the use of RAII (§6.3). Use of resource handles, such as **scoped_lock** and **unique_lock** (§18.4), is simpler and far safer than explicitly locking and unlocking **mutex**es.

The correspondence between the shared data and a **mutex** relies on convention: the programmer has to know which **mutex** is supposed to correspond to which data. Obviously, this is error-prone, and equally obviously we try to make the correspondence clear through various language means. For example:

```
class Record {
public:
        mutex rm;
        // ...
};
```

It doesn't take a genius to guess that for a **Record** called **rec**, you are supposed to acquire **rec.rm** before accessing the rest of **rec**, though a comment or a better name might have helped the reader.

It is not uncommon to need to simultaneously access several resources to perform some action. This can lead to deadlock. For example, if **thread1** acquires **mutex1** and then tries to acquire **mutex2** while **thread2** acquires **mutex2** and then tries to acquire **mutex1**, then neither task will ever proceed further. The **scoped_lock** helps by enabling us to acquire several locks simultaneously:

```
void f()
{
        scoped_lock lck {mutex1,mutex2,mutex3};      // acquire all three locks
        // ... manipulate shared data ...
} // implicitly release all mutexes
```

This **scoped_lock** will proceed only after acquiring all its **mutex**es arguments and will never block ("go to sleep") while holding a **mutex**. The destructor for **scoped_lock** ensures that the **mutex**es are released when a **thread** leaves the scope.

Communicating through shared data is pretty low level. In particular, the programmer has to devise ways of knowing what work has and has not been done by various tasks. In that regard, use of shared data is inferior to the notion of call and return. On the other hand, some people are convinced that sharing must be more efficient than copying arguments and returns. That can indeed be so when large amounts of data are involved, but locking and unlocking are relatively expensive operations. On the other hand, modern machines are very good at copying data, especially compact data, such as **vector** elements. So don't choose shared data for communication because of "efficiency" without thought and preferably not without measurement.

The basic **mutex** allows one thread at a time to access data. One of the most common ways of sharing data is among many readers and a single writer. This "reader-writer lock" idiom is supported by **shared_mutex**. A reader will acquire the mutex "shared" so that other readers can still gain access, whereas a writer will demand exclusive access. For example:

```
shared_mutex mx;            // a mutex that can be shared

void reader()
{
    shared_lock lck {mx};           // willing to share access with other readers
    // ... read ...
}

void writer()
{
    unique_lock lck {mx};           // needs exclusive (unique) access
    // ... write ...
}
```

18.3.2 atomics（原子量）

A **mutex** is a fairly heavyweight mechanism involving the operating system. It allows arbitrary amounts of work to be done free of data races. However, there is a far simpler and cheaper mechanism for doing just a tiny amount of work: an **atomic** variable. For example, here is a simple variant of the classic double-checked locking:

```
mutex mut;
atomic<bool> init_x;        // initially false.
X x;                        // variable that requires nontrivial initialization

if (!init_x) {
    lock_guard lck {mut};
    if (!init_x) {
        // ... do nontrivial initialization of x ...
        init_x = true;
    }
}

// ... use x ...
```

The **atomic** saves us from most uses of the much more expensive **mutex**. Had **init_x** not been **atomic**, that initialization would have failed ever so infrequently, causing mysterious and hard-to-find errors, because there would have been a data race on **init_x**.

Here, I used **lock_guard** rather than **scoped_lock** because I needed just one **mutex** so the simplest lock (**lock_guard**) was sufficent.

18.4 Waiting for Events（等待事件）

Sometimes, a **thread** needs to wait for some kind of external event, such as another **thread** completing a task or a certain amount of time having passed. The simplest “event” is simply time passing. Using the time facilities found in **<chrono>** I can write:

```
using namespace chrono;      // see §16.2.1

auto t0 = high_resolution_clock::now();
this_thread::sleep_for(milliseconds{20});
auto t1 = high_resolution_clock::now();

cout << duration_cast<nanoseconds>(t1-t0).count() << " nanoseconds passed\n";
```

I didn't even have to launch a **thread**; by default, **this_thread** can refer to the one and only thread.

I used **duration_cast** to adjust the clock's units to the nanoseconds I wanted.

The basic support for communicating using external events is provided by **condition_variable**s found in **<condition_variable>**. A **condition_variable** is a mechanism allowing one **thread** to wait for another. In particular, it allows a **thread** to wait for some *condition* (often called an *event*) to occur as the result of work done by other **threads**.

Using **condition_variable**s supports many forms of elegant and efficient sharing but can be rather tricky. Consider the classic example of two **thread**s communicating by passing messages through a **queue**. For simplicity, I declare the **queue** and the mechanism for avoiding race conditions on that **queue** global to the producer and consumer:

```
class Message {      // object to be communicated
    // ...
};

queue<Message> mqueue;              // the queue of messages
condition_variable mcond;           // the variable communicating events
mutex mmutex;                       // for synchronizing access to mcond
```

The types **queue**, **condition_variable**, and **mutex** are provided by the standard library.

The **consumer()** reads and processes **Message**s:

```
void consumer()
{
    while(true) {
        unique_lock lck {mmutex};                              // acquire mmutex
        mcond.wait(lck,[] { return !mqueue.empty(); });        // release mmutex and wait;
                                                               // re-acquire mmutex upon wakeup
                                                               // don't wake up unless mqueue is non-empty
        auto m = mqueue.front();                               // get the message
        mqueue.pop();
        lck.unlock();                                          // release mmutex
        // ... process m ...
    }
}
```

Here, I explicitly protect the operations on the **queue** and on the **condition_variable** with a **unique_lock** on the **mutex**. Waiting on a **condition_variable** releases its lock argument until the wait is over (so that the queue is non-empty) and then reacquires it. The explicit check of the condition, here **!mqueue.empty()**, protects against waking up just to find that some other task has "gotten there first" so that the condition no longer holds.

I used a **unique_lock** rather than a **scoped_lock** for two reasons:

- We need to pass the lock to the **condition_variable**'s **wait()**. A **scoped_lock** cannot be moved, but a **unique_lock** can be.
- We want to unlock the **mutex** protecting the condition variable before processing the message. A **unique_lock** offers operations, such as **lock()** and **unlock()**, for low-level control of synchronization.

On the other hand, **unique_lock** can handle only a single **mutex**.

The corresponding **producer** looks like this:

```
void producer()
{
      while(true) {
            Message m;
            // ... fill the message ...
            scoped_lock lck {mmutex};         // protect operations
            mqueue.push(m);
            mcond.notify_one();               // notify
      }                                       // release mmutex (at end of scope)
}
```

18.5 Communicating Tasks（任务间通信）

The standard library provides a few facilities to allow programmers to operate at the conceptual level of tasks (work to potentially be done concurrently) rather than directly at the lower level of threads and locks:

- **future** and **promise** for returning a value from a task spawned on a separate thread
- **packaged_task** to help launch tasks and connect up the mechanisms for returning a result
- **async()** for launching of a task in a manner very similar to calling a function

These facilities are found in **<future>**.

18.5.1 future and promise（future 和 promise）

The important point about **future** and **promise** is that they enable a transfer of a value between two tasks without explicit use of a lock; "the system" implements the transfer efficiently. The basic idea is simple: when a task wants to pass a value to another, it puts the value into a **promise**. Somehow, the implementation makes that value appear in the corresponding **future**, from which it can be read (typically by the launcher of the task). We can represent this graphically:

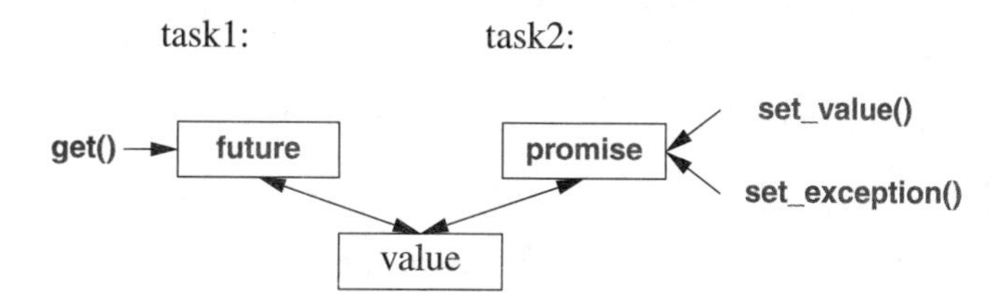

If we have a **future<X>** called **fx**, we can **get()** a value of type **X** from it:

```
X v = fx.get();   // if necessary, wait for the value to get computed
```

If the value isn't there yet, our thread is blocked until it arrives. If the value couldn't be computed, **get()** might throw an exception (from the system or transmitted from the **promise**).

The main purpose of a **promise** is to provide simple "put" operations (called **set_value()** and **set_exception()**) to match **future**'s **get()**. The names "future" and "promise" are historical; please don't blame or credit me. They are yet another fertile source of puns.

If you have a **promise** and need to send a result of type **X** to a **future**, you can do one of two things: pass a value or pass an exception. For example:

```
void f(promise<X>& px)  // a task: place the result in px
{
	// ...
	try {
		X res;
		// ... compute a value for res ...
		px.set_value(res);
	}
	catch (...) {		// oops: couldn't compute res
		px.set_exception(current_exception());		// pass the exception to the future's thread
	}
}
```

The **current_exception()** refers to the caught exception.

To deal with an exception transmitted through a **future**, the caller of **get()** must be prepared to catch it somewhere. For example:

```
void g(future<X>& fx)		// a task: get the result from fx
{
	// ...
	try {
		X v = fx.get();  // if necessary, wait for the value to get computed
		// ... use v ...
	}
	catch (...) {		// oops: someone couldn't compute v
		// ... handle error ...
	}
}
```

If the error doesn't need to be handled by **g()** itself, the code reduces to the minimal:

```
void g(future<X>& fx)		// a task: get the result from fx
{
	// ...
	X v = fx.get();  // if necessary, wait for the value to be computed
	// ... use v ...
}
```

Now, an exception thrown from **fx**'s function (**f()**) is implicitly propagated to **g()**'s caller, exactly as it would have been had **g()** called **f()** directly.

18.5.2 packaged_task

How do we get a **future** into the task that need a result and the corresponding **promise** into the thread that should produce that result? The **packaged_task** type is provided to simplify setting up tasks connected with **futures** and **promises** to be run on **threads**. A **packaged_task** provides wrapper code to put the return value or exception from the task into a **promise** (like the code shown in §18.5.1). If you ask it by calling **get_future**, a **packaged_task** will give you the **future** corresponding to its **promise**. For example, we can set up two tasks to each add half of the elements of a **vector<double>** using the standard-library **accumulate()** (§17.3):

```
double accum(vector<double>::iterator beg, vector<double>::iterator end, double init)
     // compute the sum of [beg:end) starting with the initial value init
{
     return accumulate(&*beg,&*end,init);
}

double comp2(vector<double>& v)
{
     packaged_task pt0 {accum};                              // package the task (i.e., accum)
     packaged_task pt1 {accum};

     future<double> f0 {pt0.get_future()};                   // get hold of pt0's future
     future<double> f1 {pt1.get_future()};                   // get hold of pt1's future

     double* first = &v[0];
     thread t1 {move(pt0),first,first+v.size()/2,0};          // start a thread for pt0
     thread t2 {move(pt1),first+v.size()/2,first+v.size(),0}; // start a thread for pt1
     // ...

     return f0.get()+f1.get();                               // get the results
}
```

The **packaged_task** template takes the type of the task as its template argument (here, **double(double*,double*,double)**) and the task as its constructor argument (here, **accum**). The **move()** operations are needed because a **packaged_task** cannot be copied. The reason that a **packaged_task** cannot be copied is that it is a resource handle: it owns its **promise** and is (indirectly) responsible for whatever resources its task may own.

Please note the absence of explicit mention of locks in this code: we are able to concentrate on tasks to be done, rather than on the mechanisms used to manage their communication. The two tasks will be run on separate threads and thus potentially in parallel.

18.5.3 async()

The line of thinking I have pursued in this chapter is the one I believe to be the simplest yet still among the most powerful: treat a task as a function that may happen to run concurrently with other tasks. It is far from the only model supported by the C++ standard library, but it serves well for a wide range of needs. More subtle and tricky models (e.g., styles of programming relying on shared memory), can be used as needed.

To launch tasks to potentially run asynchronously, we can use **async()**:

```
double comp4(vector<double>& v)
     // spawn many tasks if v is large enough
{
     if (v.size()<10'000)          // is it worth using concurrency?
          return accum(v.begin(),v.end(),0.0);

     auto v0 = &v[0];
     auto sz = v.size();

     auto f0 = async(accum,v0,v0+sz/4,0.0);              // first quarter
     auto f1 = async(accum,v0+sz/4,v0+sz/2,0.0);         // second quarter
     auto f2 = async(accum,v0+sz/2,v0+sz*3/4,0.0);       // third quarter
     auto f3 = async(accum,v0+sz*3/4,v0+sz,0.0);         // fourth quarter

     return f0.get()+f1.get()+f2.get()+f3.get(); // collect and combine the results
}
```

Basically, **async()** separates the "call part" of a function call from the "get the result part" and separates both from the actual execution of the task. Using **async()**, you don't have to think about threads and locks. Instead, you think in terms of tasks that potentially compute their results asynchronously. There is an obvious limitation: don't even think of using **async()** for tasks that share resources needing locking. With **async()** you don't even know how many **thread**s will be used because that's up to **async()** to decide based on what it knows about the system resources available at the time of a call. For example, **async()** may check whether any idle cores (processors) are available before deciding how many **thread**s to use.

Using a guess about the cost of computation relative to the cost of launching a **thread**, such as **v.size()<10'000**, is very primitive and prone to gross mistakes about performance. However, this is not the place for a proper discussion about how to manage **thread**s. Don't take this estimate as more than a simple and probably poor guess.

It is rarely necessary to manually parallelize a standard-library algorithm, such as **accumulate()**, because the parallel algorithms (e.g., **reduce(par_unseq,/*...*/)**) usually do a better job at that (§17.3.1). However, the technique is general.

Please note that **async()** is not just a mechanism specialized for parallel computation for increased performance. For example, it can also be used to spawn a task for getting information from a user, leaving the "main program" active with something else (§18.5.3).

18.5.4 Stopping a thread（停止 thread）

Sometimes, we want to stop a **thread** because we no longer are interested in its result. Just "killing" it is usually not acceptable because a **thread** can own resources that must be released (e.g., locks, sub-threads, and database connections). Instead, the standard library provides a mechanism for politely requesting a **thread** to clean up and go away: a **stop_token**. A **thread** can be programmed to terminate if it has a **stop_token** and is requested to stop.

Consider a parallel algorithm **find_any()** that spawns many **thread**s looking for a result. When a **thread** returns with an answer, we would like to stop the remaining **thread**s. Each **thread** spawned by find_any() calls find() to do the real work. This find() is a very simple example of a common style of task where there is a main loop in which we can insert a test of whether to continue or stop:

```
atomic<int> result = -1;   // put a resulting index here

template<class T> struct Range { T* first; T* last; };      // a way of passing a range of Ts

void find(stop_token tok, const string* base, const Range<string> r, const string target)
{
     for (string* p = r.first; p!=r.last && !tok.stop_requested(); ++p)
          if (match(*p, target)) {     // match() applies some matching criteria to the two strings
               result = p - base;              // the index of the found element
      return;
     }
}
```

Here, !tok.stop_requested() tests whether some other thread has requested this thread to terminate. A stop_token is the mechanism for safely (no data races) communicating such a request.

Here is a trivial find_any() that spawns just two threads running find():

```
void find_all(vector<string>& vs, const string& key)
{
     int mid = vs.size()/2;
     string* pvs = &vs[0];

     stop_source ss1{};
     jthread t1(find, ss1.get_token(), pvs, Range{pvs,pvs+mid}, key);                // first half of vs

     stop_source ss2{};
     jthread t2(find, ss2.get_token(), pvs, Range{pvs+mid,pvs+vs.size()} , key);  // second half of vs

     while (result == -1)
          this_thread::sleep_for(10ms);

     ss1.request_stop(); // we have a result: stop all threads
     ss2.request_stop();

     // ... use result ...
}
```

The stop_sources produces the stop_tokens through which requests to stop are communicated to threads.

The synchronization and returning of a result is the simplest I could think of: put the result in an atomic variable (§18.3.2) and do a spin loop on that.

Of course, we could elaborate this simple example to use many searcher threads, make the return of results more general, and use different element types. However, that would obscure the basic role of the stop_source and stop_token.

18.6 Coroutines（协程）

A coroutine is a function that maintains its state between calls. In that, it's a bit like a function object, but the saving and restoring of its state between calls are implicit and complete. Consider a classic example:

```
generator<long long> fib()     // generate Fibonacci numbers
{
        long long a = 0;
        long long b = 1;
        while (a<b) {
                auto next = a+b;
                co_yield next;          // save state, return value, and wait
                a = b;
                b = next;
        }
        co_return 0;                    // a fib too far
}

void user(int max)
{
        for (int i=0; i++<max;)
                cout << fib() << ' ';
}
```

This generates

```
1 2 3 5 8 13 ...
```

The **generator** return value is where the coroutine stores its state between calls. We could, of course, have made a function object **Fib** that worked the same way, but then we would have had to maintain its state ourselves. For larger states and more complex computations, saving and restoring state get tedious, hard to optimize, and error-prone. In effect, a coroutine is a function that saves its stackframe between calls. The **co_yield** returns a value and waits for the next call. The **co_return** returns a value and terminates the coroutine.

Coroutines can be synchronous (the caller waits for the result) or asynchronous (the caller does some other work until it looks for the result from the coroutine). The Fibonacci example was obviously synchronous. This allows some nice optimizations. For example, a good optimizer can inline the calls to **fib()** and unroll the loop, leaving just a sequence of calls of <<, which themselves can be optimized to

```
cout << "1 2 3 5 7 12";         // fib(6)
```

The coroutines are implemented as an extremely flexible framework able to serve an extreme range of potential uses. It is designed by and for experts with a touch of design by committee. That's fine except that the library facilities to make simple uses simple are still missing in C++20. For example, **generator** is not (yet) part of the standard library. There are proposals, though, and a Web search will find you good implementations; [Cppcoro] is an example.

18.6.1 Cooperative Multitasking（协作式多任务）

In the first volume of *The Art of Computer Programming*, Donald Knuth praises the usefulness of coroutines, but also bemoans that it is hard to give brief examples because coroutines are most useful in simplifying complex systems. Here, I will just give a trivial example exercising the primitives needed for the kind of event-driven simulations that were a major reason for C++'s early success. The key idea is to represent a system as a network of simple tasks (coroutines) that collaborate to complete complex tasks. Basically, each task is an actor that performs a small part of a large effort. Some are generators that produce streams of requests (possibly using random number generators, possibly feeding real-world data), some are parts of a network computing results, and some generate output. I personally prefer the tasks (coroutines) to communicate through message queues. One way to organize such a system is for each task to place itself on an event queue waiting for more work after producing a result. A scheduler then picks the next task to run from the event queue when needed. This is a form of *cooperative multitasking*. I borrowed – with acknowledgments – the key ideas for that from Simula [Dahl,1970] to form the basis for the very first C++ library (§19.1.2).

The keys to such designs are

- Many different *coroutines* that maintain their state between calls.
- A form of *polymorphism* that allows us to keep lists of events containing different kinds of coroutines and invoke them independently of their types.
- A *scheduler* that selects the next coroutine(s) to run from the list(s).

Here, I will just show a couple of coroutines and execute them alternatively. It is essential for such systems not to use too much space. That's why we don't use processes or threads for such applications. A thread requires a megabyte or two (mostly for its stack), a coroutine often only a couple of dozen bytes. If you need many thousands of tasks, that can make a big difference. Context switching between coroutines is also far faster than between threads or processes.

First, we need some run-time polymorphism to allow us to call dozens or hundreds of different kinds of coroutines uniformly:

```
struct Event_base {
      virtual void operator()() = 0;
      virtual ~Event_base() {}
};

template<class Act>
struct Event : Event_base {
      Event(const string n, Act a) : name{ n }, act{ move(a) } {}
      string name;
      Act act;
      void operator()() override { act(); }
};
```

An **Event** simply stores an action and allows it to be called; The action will typically be a coroutine. I added a **name** just to illustrate that an event typically carries more information than just the coroutine handle.

Here is a trivial use:

```
void test()
{
	vector<Event_base*> events = {				// create a couple of Events holding coroutines
		new Event{ "integers ", sequencer(10) },
		new Event{ "chars    ", char_seq('a') }
	};

	vector order {0, 1, 1, 0, 1, 0, 1, 0, 0};		// choose some order

	for (int x : order)					// invoke coroutines in order
		(*events[x])();

	for (auto p : events)					// clean up
		delete p;
}
```

So far, there is nothing specifically coroutine about this; it's just a conventional object-oriented framework for executing operations on a set of objects of potentially differing types. Hoewever, **sequece** and **char_seq** happen to be coroutines. The fact that they maintain their state between calls is essential for real-world uses of such frameworks:

```
task sequencer(int start, int strp =1)
{
	auto value = start;
	while (true) {
		cout << "value: " << value << '\n';		// communicate a result
		co_yield 0;					// sleep until someone resumes this coroutine
		value += step;					// update state
	}
}
```

We can see that **sequencer** is a coroutine because it used **co_yield** to suspend itself between calls. This implies that **task** must be a coroutine handle (see below).

This is a deliberately trivial coroutine. All it does is generate a sequence of values and output them. In a serious simulation, that output would directly or indirectly become the input to some other coroutine.

The **char_seq** is very similar, but of a different type to exercise the run-time polymorphism:

```
task char_seq(char start)
{
	auto value = start;
	while (true) {
		cout << "value: " << value << '\n';	// communicate result
		co_yield 0;
		++value;
	}
}
```

The "magic" is in the return type **task**; it holds the state of the coroutine (in effect the function's stack frame) between calls and determines the meaning of the **co_yield**. From a user's point of view

task is trivial, it simply provides an operator to invoke the coroutine:

```
struct task {
    void operator()();
    // ... implementation details ...
}
```

If **task** had been in a library, preferably the standard library, that would be all we needed to know, but it is not, so here is a hint of how to implement such coroutine-handle types. There are proposals, though, and a Web search will find you good implementations; the [Cppcoro] library is an example.

My **task** is as minimal I could think of to implement my key examples:

```
struct task {
    void operator()() { coro.resume(); }

    struct promise_type {                                        // mapping to the language features
        suspend_always initial_suspend() { return {}; }
        suspend_always final_suspend() noexcept { return {}; }      // co_return
        suspend_always yield_value(int) {  return {}; }             // co_yield
        auto get_return_object() { return task{ handle_type::from_promise(*this) }; }
        void return_void() {}
        void unhandled_exception() { exit(1); }
    };

    using handle_type = coroutine_handle<promise_type>;
    task(handle_type h) : coro(h) { }     // called by get_return_object()
    handle_type coro;                     // here is the coroutine handle
};
```

I strongly encourage you *not* to write such code yourself unless you are a library implementer trying to save others from the bother. If you are curious, there are many explanations on the Web.

18.7 Advice（建议）

[1] Use concurrency to improve responsiveness or to improve throughput; §18.1.
[2] Work at the highest level of abstraction that you can afford; §18.1.
[3] Consider processes as an alternative to threads; §18.1.
[4] The standard-library concurrency facilities are type safe; §18.1.
[5] The memory model exists to save most programmers from having to think about the machine architecture level of computers; §18.1.
[6] The memory model makes memory appear roughly as naively expected; §18.1.
[7] Atomics allow for lock-free programming; §18.1.
[8] Leave lock-free programming to experts; §18.1.
[9] Sometimes, a sequential solution is simpler and faster than a concurrent solution; §18.1.
[10] Avoid data races; §18.1, §18.2.
[11] Prefer parallel algorithms to direct use of concurrency; §18.1, §18.5.3.

[12] A thread is a type-safe interface to a system thread; §18.2.
[13] Use join() to wait for a thread to complete; §18.2.
[14] Prefer jthread over thread; §18.2.
[15] Avoid explicitly shared data whenever you can; §18.2.
[16] Prefer RAII to explicit lock/unlock; §18.3; [CG: CP.20].
[17] Use scoped_lock to manage mutexes; §18.3.
[18] Use scoped_lock to acquire multiple locks; §18.3; [CG: CP.21].
[19] Use shared_lock to implement reader-write locks; §18.3.
[20] Define a mutex together with the data it protects; §18.3; [CG: CP.50].
[21] Use atomics for very simple sharing; §18.3.2.
[22] Use condition_variables to manage communication among threads; §18.4.
[23] Use unique_lock (rather than scoped_lock) when you need to copy a lock or need lower-level manipulation of synchronization; §18.4.
[24] Use unique_lock (rather than scoped_lock) with condition_variables; §18.4.
[25] Don't wait without a condition; §18.4; [CG: CP.42].
[26] Minimize time spent in a critical section; §18.4 [CG: CP.43].
[27] Think in terms of concurrent tasks, rather than directly in terms of threads; §18.5.
[28] Value simplicity; §18.5.
[29] Prefer packaged_tasks and futures over direct use of threads and mutexes; §18.5.
[30] Return a result using a promise and get a result from a future; §18.5.1; [CG: CP.60].
[31] Use packaged_tasks to handle exceptions thrown by tasks; §18.5.2.
[32] Use a packaged_task and a future to express a request to an external service and wait for its response; §18.5.2.
[33] Use async() to launch simple tasks; §18.5.3; [CG: CP.61].
[34] Use stop_token to implement cooperative termination; §18.5.4.
[35] A coroutine can be very much smaller than a thread; §18.6.
[36] Prefer coroutine support libraries to hand-crafted code; §18.6.

19

History and Compatibility

（历史和兼容性）

Hurry Slowly
(festina lente).
– Octavius, Caesar Augustus

- History
  Timeline; The Early Years; The ISO C++ Standards; Standards and Programming Style; C++ Use; The C++ Model
- C++ Feature Evolution
  C++11 Language Features; C++14 Language Features; C++17 Language Features; C++20 Language Features; C++11 Standard-Library Components; C++14 Standard-Library Components; C++17 Standard-Library Components; C++20 Standard-Library Components; Removed and Deprecated Features
- C/C++ Compatibility
  C and C++ Are Siblings; Compatibility Problems
- Bibliography
- Advice

19.1 History（历史）

I invented C++, wrote its early definitions, and produced its first implementation. I chose and formulated the design criteria for C++, designed its major language features, developed or helped to develop many of the early libraries, and for 25 years was responsible for the processing of extension proposals in the C++ standards committee.

C++ was designed to provide Simula's facilities for program organization [Dahl,1970] together with C's efficiency and flexibility for systems programming [Kernighan,1978]. Simula was the initial source of C++'s abstraction mechanisms. The notion of class (with derived classes and virtual functions) was borrowed from it. However, templates and exceptions came to C++ later with different sources of inspiration.

The evolution of C++ was always in the context of its use. I spent a lot of time listening to users, seeking out the opinions of experienced programmers, and of course writing code. In particular, my colleagues at AT&T Bell Laboratories were essential for the growth of C++ during its first decade.

This section is a brief overview; it does not try to mention every language feature and library component. Furthermore, it does not go into details. For more information, and in particular for more names of people who contributed, see my three papers from the ACM History of Programming Languages conferences [Stroustrup,1993] [Stroustrup,2007] [Stroustrup,2020] and my *Design and Evolution of C++* book (known as "D&E") [Stroustrup,1994]. They describe the design and evolution of C++ in detail and document influences from and on other programming languages. I try to maintain a connection between the standard facilities and the people who proposed and refined those facilities. C++ is not the work of a faceless, anonymous committee or a supposedly omnipotent "dictator for life"; it is the work of many dedicated, experienced, hardworking individuals.

Most of the documents produced as part of the ISO C++ standards effort are available online [WG21].

19.1.1 Timeline（大事年表）

The work that led to C++ started in the fall of 1979 under the name "C with Classes." Here is a simplified timeline:

1979 Work on "C with Classes" started. The initial feature set included classes and derived classes, public/private access control, constructors and destructors, and function declarations with argument checking. The first library supported non-preemptive concurrent tasks and random number generators.

1984 "C with Classes" was renamed to C++. By then, C++ had acquired virtual functions, function and operator overloading, references, and the I/O stream and complex number libraries.

1985 First commercial release of C++ (October 14). The library included I/O streams, complex numbers, and tasks (non-preemptive scheduling).

1985 *The C++ Programming Language* ("TC++PL," October 14) [Stroustrup,1986].

1989 *The Annotated C++ Reference Manual* ("the ARM") [Ellis,1989].

1991 *The C++ Programming Language, Second Edition* [Stroustrup,1991], presenting generic programming using templates and error handling based on exceptions, including the "Resource Acquisition Is Initialization" (RAII) general resource-management idiom.

1997 *The C++ Programming Language, Third Edition* [Stroustrup,1997] introduced ISO C++, including namespaces, **dynamic_cast**, and many refinements of templates. The standard library added the STL framework of generic containers and algorithms.

1998 ISO C++ standard [C++,1998].

2002 Work on a revised standard, colloquially named C++0x, started.

2003 A "bug fix" revision of the ISO C++ standard was issued. [C++,2011].

2011 ISO C++11 standard [C++,2011] offering uniform initialization, move semantics, types deduced from initializers (**auto**), range-**for**, variadic templates, lambda expressions, type aliases, a memory model suitable for concurrency, and much more. The standard library added threads, locks, regular expressions, hash tables (unordered_maps), resource-management pointers (unique_ptr and shared_ptr), and more.

2013 The first complete C++11 implementations emerged.

2013 *The C++ Programming Language, Fourth Edition* introduced C++11.

2014 ISO C++14 standard [C++,2014] completing C++11 with variable templates, digit separators, generic lambdas, and a few standard-library improvements. The first C++14 implementations were completed.

2015 The C++ Core Guidelines projects started [Stroustrup,2015].

2017 ISO C++17 standard [C++,2017] offering a diverse set of new features, including order of evaluation guarantees, structured bindings, fold expressions, a file system library, parallel algorithms, and variant and optional types. The first C++17 implementations were completed.

2020 ISO C++20 standard [C++,2020] offering modules, concepts, coroutines, ranges, printf()-style formatting, calendars, and many minor features. The first C++20 implementations were completed.

During development, C++11 was known as C++0x. As is not uncommon in large projects, we were overly optimistic about the completion date. Towards the end, we joked that the 'x' in C++0x was hexadecimal so that C++0x became C++0B. On the other hand, the committee shipped C++14, C++17, and C++20 on time, as did the major compiler providers.

19.1.2 The Early Years（早期的 C++）

I originally designed and implemented the language because I wanted to distribute the services of a UNIX kernel across multiprocessors and local-area networks (what are now known as multicores and clusters). For that, I needed to precisely specify parts of a system and how they communicated. Simula [Dahl,1970] would have been ideal for that, except for performance considerations. I also needed to deal directly with hardware and provide high-performance concurrent programming mechanisms for which C would have been ideal, except for its weak support for modularity and type checking. The result of adding Simula-style classes to C (Classic C; §19.3.1), "C with Classes," was used for major projects in which its facilities for writing programs that use minimal time and space were severely tested. It lacked operator overloading, references, virtual functions, templates, exceptions, and many, many details [Stroustrup,1982]. The first use of C++ outside a research organization started in July 1983.

The name C++ (pronounced "see plus plus") was coined by Rick Mascitti in the summer of 1983 and chosen as the replacement for "C with Classes" by me. The name signifies the evolutionary nature of the changes from C; "++" is the C increment operator. The slightly shorter name "C+" is a syntax error; it had also been used as the name of an unrelated language. Connoisseurs of C semantics find C++ inferior to ++C. The language was not called D, because it was an extension of C, because it did not attempt to remedy problems by removing features, and because there already existed several would-be C successors named D. For yet another interpretation of the name C++, see the appendix of [Orwell,1949].

C++ was designed primarily so that my friends and I would not have to program in assembler, C, or various then-fashionable high-level languages. Its main purpose was to make writing good programs easier and more pleasant for the individual programmer. In the early years, there was no

C++ paper design; design, documentation, and implementation went on simultaneously. There was no "C++ project" either, nor a "C++ design committee." Throughout, C++ evolved to cope with problems encountered by users and as a result of discussions among my friends, my colleagues, and me.

The very first design of C++ included function declarations with argument type checking and implicit conversions, classes with the **public/private** distinction between the interface and the implementation, derived classes, and constructors and destructors. I used macros to provide primitive parameterization [Stroustrup,1982]. This was in non-experimental use by mid-1980. Late that year, I was able to present a set of language facilities supporting a coherent set of programming styles. In retrospect, I consider the introduction of constructors and destructors most significant. In the terminology of the time [Stroustrup,1979]:

A "new function" creates the execution environment for the member functions;
the "delete function" reverses that.

Soon after, "new function" and "delete function" were renamed "constructor" and "destructor." Here is the root of C++'s strategies for resource management (causing a demand for exceptions) and the key to many techniques for making user code short and clear. If there were other languages at the time that supported multiple constructors capable of executing general code, I didn't (and don't) know of them. Destructors were new in C++.

C++ was released commercially in October 1985. By then, I had added inlining (§1.3, §5.2.1), **consts** (§1.6), function overloading (§1.3), references (§1.7), operator overloading (§5.2.1, §6.4), and virtual functions (§5.4). Of these features, support for run-time polymorphism in the form of virtual functions was by far the most controversial. I knew its worth from Simula but found it impossible to convince most people in the systems programming world of its value. Systems programmers tended to view indirect function calls with suspicion, and people acquainted with other languages supporting object-oriented programming had a hard time believing that **virtual** functions could be fast enough to be useful in systems code. Conversely, many programmers with an object-oriented background had (and many still have) a hard time getting used to the idea that you use virtual function calls only to express a choice that must be made at run time. The resistance to virtual functions may be related to a resistance to the idea that you can get better systems through more regular structure of code supported by a programming language. Many C programmers seem convinced that what really matters is complete flexibility and careful individual crafting of every detail of a program. My view was (and is) that we need every bit of help we can get from languages and tools: the inherent complexity of the systems we are trying to build is always at the edge of what we can express.

Early documents (e.g., [Stroustrup,1985] and [Stroustrup,1994]) described C++ like this:

C++ is a general-purpose programming language that

- *is a better C*
- *supports data abstraction*
- *supports object-oriented programming*

Note that I did *not* say "C++ is an object-oriented programming language." Here, "supports data abstraction" refers to information hiding, classes that are not part of class hierarchies, and generic programming. Initially, generic programming was poorly supported through the use of macros [Stroustrup,1982]. Templates and concepts came much later.

Much of the design of C++ was done on the blackboards of my colleagues. In the early years, the feedback from Stu Feldman, Alexander Fraser, Steve Johnson, Brian Kernighan, Doug McIlroy, and Dennis Ritchie was invaluable.

In the second half of the 1980s, I continued to add language features in response to user comments and guided by my general aims for C++. The most important of those were templates [Stroustrup,1988] and exception handling [Koenig,1990], which were considered experimental at the time the standards effort started. In the design of templates, I was forced to decide between flexibility, efficiency, and early type checking. At the time, nobody knew how to simultaneously get all three. To compete with C-style code for demanding systems applications, I felt that I had to choose the first two properties. In retrospect, I think the choice was the correct one, and the continued search for better type checking of templates [DosReis,2006] [Gregor,2006] [Sutton,2011] [Stroustrup,2012a] [Stroustrup,2017] led to the C++20 concepts (Chapter 8). The design of exceptions focused on multilevel propagation of exceptions, the passing of arbitrary information to an error handler, and the integration between exceptions and resource management by using local objects with destructors to represent and release resources. I clumsily named that critical technique *Resource Acquisition Is Initialization* and others soon reduced that to the acronym *RAII* (§6.3).

I generalized C++'s inheritance mechanisms to support multiple base classes [Stroustrup,1987]. This was called *multiple inheritance* and was considered difficult and controversial. I considered it far less important than templates or exceptions. Multiple inheritance of abstract classes (often called *interfaces*) is now universal in languages supporting static type checking and object-oriented programming.

The C++ language evolved hand-in-hand with some of the key library facilities. For example, I designed the complex [Stroustrup,1984], vector, stack, and (I/O) stream classes [Stroustrup,1985] together with the operator overloading mechanisms. The first string and list classes were developed by Jonathan Shopiro and me as part of the same effort. Jonathan's string and list classes were the first to see extensive use as part of a library. The string class from the standard C++ library has its roots in these early efforts. The task library described in [Stroustrup,1987b] was part of the first "C with Classes" program ever written in 1980. It provided coroutines and a scheduler. I wrote it and its associated classes to support Simula-style simulations. It was crucial for C++'s success and wide adoption in the 1980s. Unfortunately, we had to wait until 2011 (30 years!) to get concurrency support standardized and universally available (§18.6). Coroutines are part of C++20 (§18.6). The development of the template facility was influenced by a variety of **vector**, **map**, **list**, and **sort** templates devised by Andrew Koenig, Alex Stepanov, me, and others.

The most important innovation in the 1998 standard library was the STL, a framework of algorithms and containers (Chapter 12, Chapter 13). It was the work of Alex Stepanov (with Dave Musser, Meng Lee, and others) based on more than a decade's work on generic programming. The STL has been massively influential within the C++ community and beyond.

C++ grew up in an environment with a multitude of established and experimental programming languages (e.g., Ada [Ichbiah,1979], Algol 68 [Woodward,1974], and ML [Paulson,1996]). At the time, I was comfortable in about 25 languages, and their influences on C++ are documented in [Stroustrup,1994] and [Stroustrup,2007]. However, the determining influences always came from the applications I encountered. It was my deliberate policy to have the evolution of C++ "problem driven" rather than imitative.

19.1.3 The ISO C++ Standards（ISO C++ 标准）

The explosive growth of C++ use caused some changes. Sometime during 1987, it became clear that formal standardization of C++ was inevitable and that we needed to lay the groundwork for a standardization effort [Stroustrup,1994]. The result was a conscious effort to maintain contact between implementers of C++ compilers and their major users. This was done through paper and electronic mail and through face-to-face meetings at C++ conferences and elsewhere.

AT&T Bell Labs made a major contribution to C++ and its wider community by allowing me to share drafts of revised versions of the C++ reference manual with implementers and users. Because many of those people worked for companies that could be seen as competing with AT&T, the significance of this contribution should not be underestimated. A less enlightened company could have caused major problems of language fragmentation simply by doing nothing. As it happened, about a hundred individuals from dozens of organizations read and commented on what became the generally accepted reference manual and the base document for the ANSI C++ standardization effort. Their names can be found in *The Annotated C++ Reference Manual* ("the ARM") [Ellis,1989]. The X3J16 committee of ANSI was convened in December 1989 at the initiative of Hewlett-Packard, DEC, and IBM with the support of AT&T. In June 1991, this ANSI (American national) standardization of C++ became part of an ISO (international) standardization effort for C++. The ISO C++ committee is called WG21. From 1990 onward, these joint C++ standards committees have been the main forum for the evolution of C++ and the refinement of its definition. I served on these committees throughout. In particular, as the chairman of the working group for extensions (later called the evolution group) from 1990 to 2014, I was directly responsible for handling proposals for major changes to C++ and the addition of new language features. An initial draft standard for public review was produced in April 1995. The first ISO C++ standard (ISO/IEC 14882-1998) [C++,1998] was ratified by a 22-0 national vote in 1998. A "bug fix release" of this standard was issued in 2003, so you sometimes hear people refer to C++03, but that is essentially the same language and standard library as C++98.

C++11, known for years as C++0x, is the work of the members of WG21. The committee worked under increasingly onerous self-imposed processes and procedures. These processes probably led to a better (and more rigorous) specification, but they also limited innovation [Stroustrup,2007]. An initial draft standard for public review was produced in 2009. The second ISO C++ standard (ISO/IEC 14882-2011) [C++,2011] was ratified by a 21-0 national vote in August 2011.

One reason for the long gap between the two standards is that most members of the committee (including me) were under the mistaken impression that the ISO rules required a "waiting period" after a standard was issued before starting work on new features. Consequently, serious work on new language features did not start until 2002. Other reasons included the increased size of modern languages and their foundation libraries. In terms of pages of standards text, the language grew by about 30% and the standard library by about 100%. Much of the increase was due to more detailed specification, rather than new functionality. Also, the work on a new C++ standard obviously had to take great care not to compromise older code through incompatible changes. There are billions of lines of C++ code in use that the committee must not break. Stability over decades is an essential "feature."

C++11 added massively to the standard library and pushed to complete the feature set needed for a programming style that is a synthesis of the "paradigms" and idioms that had proven successful with C++98.

The overall aims for the C++11 effort were:

- Make C++ a better language for systems programming and library building.
- Make C++ easier to teach and learn.

The aims are documented and detailed in [Stroustrup,2007].

A major effort was made to make concurrent systems programming type-safe and portable. This involved a memory model (§18.1) and support for lock-free programming, This was the work of Hans Boehm, Brian McKnight, and others in the concurrency working group. On top of that, we added the **threads** library.

After C++11, there was wide agreement that 13 years between standards were far too many. Herb Sutter proposed that the committee adopt a policy of shipping on time at fixed intervals, the "train model." I argued strongly for a short interval between standards to minimize the chance of delays because someone insisted on extra time to allow inclusion of "just one more essential feature." We agreed on an ambitious 3-year schedule with the idea that we should alternate between minor and major releases.

C++14 was deliberately a minor release aiming at "completing C++11." This reflects the reality that with a fixed release date, there will be features that we know we want, but can't deliver on time. Also, once in widespread use, gaps in the feature set will inevitably be discovered.

C++17 was meant to be a major release. By "major," I mean containing features that will change the way we think about the structure of our software and about how we design it. By this definition, C++17 was at best a medium release. It included a lot of minor extensions, but the features that would have made dramatic changes (e.g., concepts, modules, and coroutines) were either not ready or became mired in controversy and lack of design direction. As a result, C++17 includes a little bit for everyone, but nothing that would significantly change the life of a C++ programmer who had already absorbed the lessons of C++11 and C++14.

C++20 offers long-promised and much-needed major features, such as modules (§3.2.2), concepts (§8.2), coroutines (§18.6), ranges (§14.5), and many minor features. It is as major an upgrade to C++ as was C++11. It became widely available in late 2021.

The ISO C++ standards committee, SC22/WG21, now has about 350 members, out of which about 250 attended the last pre-pandemic face-to-face meeting in Prague where C++20 was approved by a unanimous 79-0 that was later ratified by a national body vote of 22-0. Getting that degree of agreement among such a large and diverse group is hard work. Dangers include "Design by committee," feature bloat, lack of consistent style, and short-sighted decisions. Making progress toward a simpler-to-use and more coherent language is very hard. The committee is aware of that and trying to counter it; see [Wong,2020]. Sometimes, we succeed but it is very hard to avoid complexity creeping in from "minor useful features," fashion, and the desire of experts to serve rare special cases directly.

19.1.4 Standards and Style（标准与编程风格）

A standard says what will work, and how. It does not say what constitutes good and effective use. There are significant differences between understanding the technical details of programming language features and using them effectively in combination with other features, libraries, and tools to produce better software. By "better" I mean "more maintainable, less error-prone, and faster."

We need to develop, popularize, and support coherent programming styles. Further, we must support the evolution of older code to these more modern, effective, and coherent styles.

With the growth of the language and its standard library, the problem of popularizing effective programming styles became critical. It is extremely difficult to make large groups of programmers depart from something that works for something better. There are still people who see C++ as a few minor additions to C and people who consider 1980s Object-Oriented programming styles based on massive class hierarchies the pinnacle of development. Many are still struggling to use modern C++ well in environments with lots of older C++ code. On the other hand, there are also many who enthusiastically overuse novel facilities. For example, some programmers are convinced that only code using massive amounts of template metaprogramming is true C++.

What is *Modern C++*? In 2015, I set out to answer this question by developing a set of coding guidelines supported by articulated rationales. I soon found that I was not alone in grappling with that problem, and together with people from many parts of the world, notably from Microsoft, Red Hat, and Facebook, started the "C++ Core Guidelines" project [Stroustrup,2015]. This is an ambitious project aiming at complete type-safety and complete resource-safety as a base for simpler, faster, safer, and more maintainable code [Stroustrup,2015b] [Stroustrup,2021]. In addition to specific coding rules with rationales, we back up the guidelines with static analysis tools and a tiny support library. I see something like that as necessary for moving the C++ community at large forward to benefit from the improvements in language features, libraries, and supporting tools.

19.1.5 C++ Use （C++ 的使用）

C++ is now a very widely used programming language. Its user population grew quickly from one in 1979 to about 400,000 in 1991; that is, the number of users doubled about every 7.5 months for more than a decade. Naturally, the growth rate slowed since that initial growth spurt, but my best estimate is that there were about 4.5 million C++ programmers in 2018 [Kazakova,2015] and maybe a million more today (2022). Much of that growth happened after 2005 when the exponential explosion of processor speed stopped so that language performance grew in importance. This growth was achieved without formal marketing or an organized user community [Stroustrup,2020].

C++ is primarily an industrial language; that is, it is more prominent in industry than in education or programming language research. It grew up in Bell Labs inspired by the varied and stringent needs of telecommunications and systems programming (including device drivers, networking, and embedded systems). From there, C++ use has spread into essentially every industry: microelectronics, Web applications and infrastructure, operating systems, financial, medical, automobile, aerospace, high-energy physics, biology, energy production, machine learning, video games, graphics, animation, virtual reality, and much more. It is primarily used where problems require C++'s combination of the ability to use hardware effectively and to manage complexity. This seems to be a continuously expanding set of applications [Stroustrup,1993] [Stroustrup,2014] [Stroustrup,2020].

19.1.6 The C++ Model（C++ 模型）

The C++ language can be summarized as a set of mutually supportive facilities:

- A static type system with equal support for built-in types and user-defined types (Chapter 1, Chapter 5, Chapter 6)

- Value and reference semantics (§1.7, §5.2, §6.2, Chapter 12, §15.2)
- Systematic and general resource management (RAII) (§6.3)
- Support for efficient object-oriented programming (§5.3, class.virtual, §5.5)
- Support for flexible and efficient generic programming (Chapter 7, Chapter 18)
- Support for compile-time programming (§1.6, Chapter 7, Chapter 8)
- Direct use of machine and operating system resources (§1.4, Chapter 18)
- Concurrency support through libraries (often implemented using intrinsics) (Chapter 18)

The standard-library components add further essential support for these high-level aims.

19.2 C++ Feature Evolution（C++特性演化）

Here, I list the language features and standard-library components that have been added to C++ for the C++11, C++14, C++17, and C++20 standards.

19.2.1 C++11 Language Features（C++11 语言特性）

Looking at a list of language features can be quite bewildering. Remember that a language feature is not meant to be used in isolation. In particular, most features that are new in C++11 make no sense in isolation from the framework provided by older features.

[1] Uniform and general initialization using {}-lists (§1.4.2, §5.2.3)
[2] Type deduction from initializer: auto (§1.4.2)
[3] Prevention of narrowing (§1.4.2)
[4] Generalized and guaranteed constant expressions: constexpr (§1.6)
[5] Range-for-statement (§1.7)
[6] Null pointer keyword: nullptr (§1.7.1)
[7] Scoped and strongly typed enums: enum class (§2.4)
[8] Compile-time assertions: static_assert (§4.5.2)
[9] Language mapping of {}-list to std::initializer_list (§5.2.3)
[10] Rvalue references, enabling move semantics (§6.2.2)
[11] Lambdas (§7.3.3)
[12] Variadic templates (§7.4.1)
[13] Type and template aliases (§7.4.2)
[14] Unicode characters
[15] long long integer type
[16] Alignment controls: alignas and alignof
[17] The ability to use the type of an expression as a type in a declaration: decltype
[18] Raw string literals (§10.4)
[19] Suffix return type syntax (§3.4.4)
[20] A syntax for attributes and two standard attributes: [[carries_dependency]] and [[noreturn]]
[21] A way of preventing exception propagation: the noexcept specifier (§4.4)
[22] Testing for the possibility of a throw in an expression: the noexcept operator
[23] C99 features: extended integral types (i.e., rules for optional longer integer types); concatenation of narrow/wide strings; __STDC_HOSTED__; _Pragma(X); vararg macros and empty macro arguments

[24] __func__ as the name of a string holding the name of the current function
[25] inline namespaces
[26] Delegating constructors
[27] In-class member initializers (§6.1.3)
[28] Control of defaults: default and delete (§6.1.1)
[29] Explicit conversion operators
[30] User-defined literals (§6.6)
[31] More explicit control of template instantiation: extern templates
[32] Default template arguments for function templates
[33] Inheriting constructors (§12.2.2)
[34] Override controls: override (§5.5) and final
[35] A simpler and more general SFINAE (Substitution Failure Is Not An Error) rule
[36] Memory model (§18.1)
[37] Thread-local storage: thread_local

For a more complete description of the changes to C++98 in C++11, see [Stroustrup,2013].

19.2.2 C++14 Language Features（C++14 语言特性）

[1] Function return-type deduction; (§3.4.3)
[2] Improved constexpr functions, e.g., for-loops allowed (§1.6)
[3] Variable templates (§7.4.1)
[4] Binary literals (§1.4)
[5] Digit separators (§1.4)
[6] Generic lambdas (§7.3.3.1)
[7] More general lambda capture
[8] [[deprecated]] attribute
[9] A few more minor extensions

19.2.3 C++17 Language Features（C++17 语言特性）

[1] Guaranteed copy elision (§6.2.2)
[2] Dynamic allocation of over-aligned types
[3] Stricter order of evaluation (§1.4.1)
[4] UTF-8 literals (u8)
[5] Hexadecimal floating-point literals (§11.6.1)
[6] Fold expressions (§8.4.1)
[7] Generic value template arguments (auto template parameters; §8.2.5)
[8] Class template argument type deduction (§7.2.3)
[9] Compile-time if (§7.4.3)
[10] Selection statements with initializers (§1.8)
[11] constexpr lambdas
[12] inline variables
[13] Structured bindings (§3.4.5)
[14] New standard attributes: [[fallthrough]], [[nodiscard]], and [[maybe_unused]]

[15] std::byte type (§16.7)
[16] Initialization of an enum by a value of its underlying type (§2.4)
[17] A few more minor extensions

19.2.4 C++20 Language Features （C++20 语言特性）

[1] Modules (§3.2.2)
[2] Concepts (§8.2)
[3] Coroutines (§18.6)
[4] Designated initializers (a slightly restricted version of a C99 feature)
[5] <=> (the "spaceship operator") a three-way comparison (§6.5.1)
[6] [*this] to capture a current object by value (§7.3.3)
[7] Standard attributes [[no_unique_address]], [[likely]], and [[unlikely]]
[8] More facilities allowed in constexpr functions, including new, union, try-catch, dynamic_cast, and typeid.
[9] consteval functions guaranteeing compile-time evaluation (§1.6)
[10] constinit variables to guarantee static (not run-time) initialization (§1.6)
[11] using scoped enums (§2.4)
[12] A few more minor extensions

19.2.5 C++11 Standard-Library Components （C++11 标准库组件）

The C++11 additions to the standard library come in two forms: new components (such as the regular expression matching library) and improvements to C++98 components (such as move constructors for containers).

[1] initializer_list constructors for containers (§5.2.3)
[2] Move semantics for containers (§6.2.2, §13.2)
[3] A singly-linked list: forward_list (§12.3)
[4] Hash containers: unordered_map, unordered_multimap, unordered_set, and unordered_multiset (§12.6, §12.8)
[5] Resource management pointers: unique_ptr, shared_ptr, and weak_ptr (§15.2.1)
[6] Concurrency support: thread (§18.2), mutexes and locks (§18.3), and condition variables (§18.4)
[7] Higher-level concurrency support: packaged_thread, future, promise, and async() (§18.5)
[8] tuples (§15.3.4)
[9] Regular expressions: regex (§10.4)
[10] Random numbers: distributions and engines (§17.5)
[11] Integer type names, such as int16_t, uint32_t, and int_fast64_t (§17.8)
[12] A fixed-sized contiguous sequence container: array (§15.3)
[13] Copying and rethrowing exceptions (§18.5.1)
[14] Error reporting using error codes: system_error
[15] emplace() operations for containers (§12.8)
[16] Wide use of constexpr functions
[17] Systematic use of noexcept functions

[18] Improved function adaptors: **function** and **bind()** (§16.3)
[19] **string** to numeric value conversions
[20] Scoped allocators
[21] Type traits, such as **is_integral** and **is_base_of** (§16.4.1)
[22] Time utilities: **duration** and **time_point** (§16.2.1)
[23] Compile-time rational arithmetic: **ratio**
[24] Abandoning a process: **quick_exit** (§16.8)
[25] More algorithms, such as **move()**, **copy_if()**, and **is_sorted()** (Chapter 13)
[26] Garbage collection API; later deprecated (§19.2.9)
[27] Low-level concurrency support: **atomics** (§18.3.2)
[28] A few more minor extensions

19.2.6 C++14 Standard-Library Components（C++14 标准库组件）

[1] **shared_mutex** and **shared_lock** (§18.3)
[2] User-defined literals (§6.6)
[3] Tuple addressing by type (§15.3.4)
[4] Associative container heterogenous lookup
[5] A few more minor extensions

19.2.7 C++17 Standard-Library Components（C++17 标准库组件）

[1] File system (§11.9)
[2] Parallel algorithms (§13.6, §17.3.1)
[3] Mathematical special functions (§17.2)
[4] **string_view** (§10.3)
[5] **any** (§15.4.3)
[6] **variant** (§15.4.1)
[7] **optional** (§15.4.2)
[8] A way of invoking anything that can be called for a given set of arguments: **invoke()**
[9] Elementary string conversions: **to_chars()** and **from_chars()**
[10] Polymorphic allocator (§12.7)
[11] **scoped_lock** (§18.3)
[12] A few more minor extensions

19.2.8 C++20 Standard-Library Components（C++20 标准库组件）

[1] Ranges, views, and pipelines (§14.1)
[2] **printf()**-style formatting: **format()** and **vformat()** (§11.6.2)
[3] Calendars (§16.2.2) and time-zones (§16.2.3)
[4] **span** for read and write access to contiguous arrays (§15.2.2)
[5] **source_location** (§16.5)
[6] Mathematical constants, e.g., **pi** and **ln10e** (§17.9)
[7] Many extensions to **atomics** (§18.3.2)

[8] Ways of waiting for a numbet of threads: barrier and latch.
[9] Feature test macros
[10] bit_cast<> (§16.7)
[11] Bit operations (§16.7)
[12] More standard-library functions made constexpr
[13] Many uses of <=> in the standard library
[14] Many more minor extensions

19.2.9 Removed and Deprecated Features （移除或弃用的特性）

There are billions of lines of C++ "out there" and nobody knows exactly what features are in critical use. Consequently, the ISO committee removes older features only reluctantly and after years of warning. However, sometimes troublesome features are removed or *deprecated*.

By deprecating a feature, the standards committee expresses the wish that the feature will go away. However, the committee does not have the power to immediately remove a heavily used feature – however redundant or dangerous it may be. Thus, a deprecation is a strong hint to avoid the feature. It may disappear in the future. The list of deprecated features is in Appendix D of the standard [C++,2020]. Compilers are likely to issue warnings for uses of deprecated features. However, deprecated features are part of the standard and history shows that they tend to remain supported "forever" for reasons of compatibility. Even features finally removed tend to live on in implementations because of user pressure on implementers.

- Removed: Exception specifications: void f() throw(X,Y); // C++98; now an error
- Removed: The support facilities for exception specifications, unexpected_handler, set_unexpected(), get_unexpected(), and unexpected(). Instead, use noexcept (§4.2).
- Removed: Trigraphs.
- Removed: auto_ptr. Instead, use unique_ptr (§15.2.1).
- Removed: The use of the storage specifier register.
- Removed: The use of ++ on a bool.
- Removed: The C++98 export feature. It was complex and not shipped by the major vendors. Instead, export is used as a keyword for modules (§3.2.2).
- Deprecated: Generation of copy operations for a class with a destructor (§6.2.1).
- Removed: Assignment of a string literal to a char*. Instead use const char* or auto.
- Removed: Some C++ standard-library function objects and associated functions. Most relate to argument binding. Instead use lambdas and function (§16.3).
- Deprecated: Comparisons of enum values with values from a different enum or a floating point value.
- Deprecated: Comparisons between two arrays.
- Deprecated: Comma operations in a subscript (e.g., [a,b]). To make room for allowing user defined operator[]() with multiple arguments.
- Deprecated: Implicit capture of *this in lambda expressions. Instead, use [=,this] (§7.3.3).
- Removed: The standard-library interface for garbage collectors. The C++ garbage collectors don't use that interface.
- Deprecated: strstream; instead, use spanstream (§11.7.4).

19.3 C/C++ Compatibility（C/C++ 兼容性）

With minor exceptions, C++ is a superset of C (meaning C11; [C,2011]). Most differences stem from C++'s greater emphasis on type checking. Well-written C programs tend to be C++ programs as well. For example, every example in K&R2 [Kernighan,1988] is C++. A compiler can diagnose every difference between C++ and C. The C11/C++20 incompatibilities are listed in Appendix C of the standard [C++,2020].

19.3.1 C and C++ Are Siblings（C 与 C++ 是兄弟）

How can I call C and C++ siblings? Look at a simplified family tree:

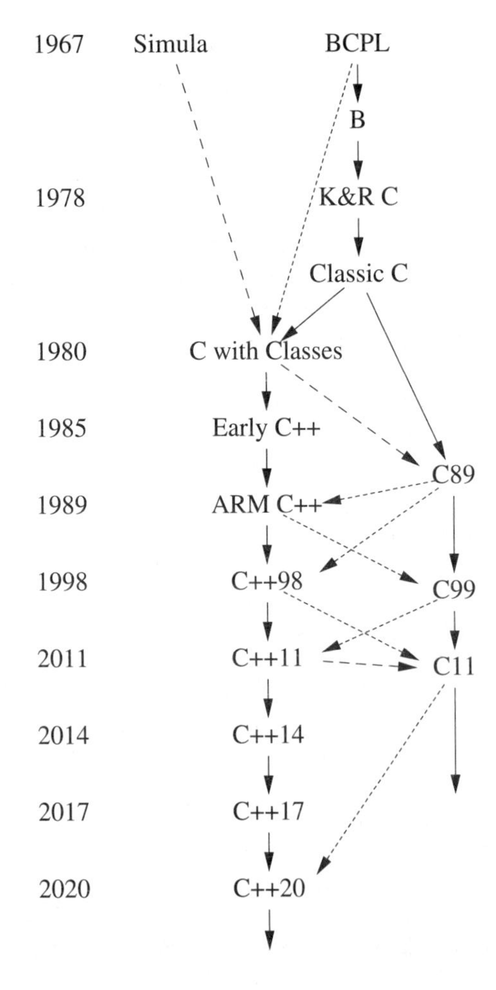

Classic C has two main descendants: ISO C and ISO C++. Over the years, these languages have evolved at different paces and in different directions. One result of this is that each language provides support for traditional C-style programming in slightly different ways. The resulting incompatibilities can make life miserable for people who use both C and C++, for people who write in one language using libraries implemented in the other, and for implementers of libraries and tools for C and C++.

A solid line means a massive inheritance of features, a dashed line a borrowing of major features, and a dotted line a borrowing of minor features. From this, ISO C and ISO C++ emerge as the two major descendants of K&R C [Kernighan,1978], and as siblings. Each carries with it the key aspects of Classic C, and neither is 100% compatible with Classic C. I picked the term "Classic C" from a sticker that used to be affixed to Dennis Ritchie's terminal. It is K&R C plus enumerations and struct assignment. BCPL is defined by [Richards,1980] and C89 by [C1990].

There was a C++03, which I didn't list because it was a bug-fix release. Similarly, C17 is not listed because it is a bug-fix release to C11.

Note that differences between C and C++ are not necessarily the result of changes to C made in C++. In several cases, the incompatibilities arise from features adopted incompatibly into C long after they were common in C++. Examples are the ability to assign a T* to a void* and the linkage of global consts [Stroustrup,2002]. Sometimes, a feature was even incompatibly adopted into C after it was part of the ISO C++ standard, such as details of the meaning of inline.

19.3.2 Compatibility Problems（兼容性问题）

There are many minor incompatibilities between C and C++. All can cause problems for a programmer but in the context of C++, all can be coped with. If nothing else, C code fragments can be compiled as C and linked to using the extern "C" mechanism.

The major problems for converting a C program to C++ are likely to be:

- Suboptimal design and programming style.
- A void* implicitly converted to a T* (that is, converted without a cast).
- C++ keywords, such as class and private, used as identifiers in C code.
- Incompatible linkage of code fragments compiled as C and fragments compiled as C++.

19.3.2.1 Style Problems（风格问题）

Naturally, a C program is written in a C style, such as the style used in K&R [Kernighan,1988]. This implies widespread use of pointers and arrays, and probably many macros. These facilities are hard to use reliably in a large program. Resource management and error handling are often ad hoc (rather than language and tool supported) and often incompletely documented and adhered to. A simple line-for-line conversion of a C program into a C++ program yields a program that is often a bit better checked. In fact, I have never converted a C program into C++ without finding some bugs. However, the fundamental structure is unchanged, and so are the fundamental sources of errors. If you had incomplete error handling, resource leaks, or buffer overflows in the original C program, they will still be there in the C++ version. To obtain major benefits, you must make changes to the fundamental structure of the code:

[1] Don't think of C++ as C with a few features added. C++ can be used that way, but only suboptimally. To get really major advantages from C++ as compared to C, you need to apply different design and implementation styles.

[2] Use the C++ standard library as a teacher of new techniques and programming styles. Note the difference from the C standard library (e.g., = rather than strcpy() for copying).
[3] Macro substitution is almost never necessary in C++. Use const (§1.6), constexpr (§1.6), enum or enum class (§2.4) to define manifest constants, constexpr (§1.6), consteval (§1.6), and inline (§5.2.1) to avoid function-calling overhead, templates (Chapter 7) to specify families of functions and types, and namespaces (§3.3) to avoid name clashes.
[4] Don't declare a variable before you need it, and initialize it immediately. A declaration can occur anywhere a statement can (§1.8), such as in for-statement initializers and in conditions (§1.8).
[5] Don't use malloc(). The new operator (§5.2.2) does the same job better, and instead of realloc(), try a vector (§6.3, §12.2). Don't just replace malloc() and free() with "naked" new and delete (§5.2.2).
[6] Avoid void*, unions, and casts, except deep within the implementation of some function or class. Their use limits the support you can get from the type system and can harm performance. In most cases, a cast is an indication of a design error.
[7] If you must use an explicit type conversion, use an appropriate named cast (e.g., static_cast; §5.2.3) for a more precise statement of what you are trying to do.
[8] Minimize the use of arrays and C-style strings. C++ standard-library strings (§10.2), arrays (§15.3.1), and vectors (§12.2) can often be used to write simpler and more maintainable code compared to the traditional C style. In general, try not to build yourself what has already been provided by the standard library.
[9] Avoid pointer arithmetic except in very specialized code (such as a memory manager).
[10] Pass contiguous sequences (e.g., arrays) as spans (§15.2.2). It's a good way to avoid range errors ("buffer overruns") without added tests.
[11] For simple array traversal, use range-for (§1.7). It's easier to write, as fast as, and safer than a traditional C loop.
[12] Use nullptr (§1.7.1) rather than 0 or NULL.
[13] Do not assume that something laboriously written in C style (avoiding C++ features such as classes, templates, and exceptions) is more efficient than a shorter alternative (e.g., using standard-library facilities). Often (but of course not always), the opposite is true.

19.3.2.2 void*

In C, a void* may be used as the right-hand operand of an assignment to or initialization of a variable of any pointer type; in C++ it may not. For example:

```
void f(int n)
{
    int* p = malloc(n*sizeof(int));   /* not C++; in C++, allocate using "new" */
    // ...
}
```

This is probably the single most difficult incompatibility to deal with. Note that the implicit conversion of a void* to a different pointer type is *not* in general harmless:

```
char ch;
void* pv = &ch;
int* pi = pv;            // not C++
*pi = 666;               // overwrite ch and other bytes near ch
```

In both languages, cast the result of **malloc()** to the right type. If you use only C++, avoid **malloc()**.

19.3.2.3 Linkage（链接）

C and C++ can be (and often are) implemented to use different linkage conventions. The most basic reason for that is C++'s greater emphasis on type checking. A practical reason is that C++ supports overloading, so there can be two global functions called **open()**. This has to be reflected in the way the linker works.

To give a C++ function C linkage (so that it can be called from a C program fragment) or to allow a C function to be called from a C++ program fragment, declare it **extern "C"**. For example:

```
extern "C" double sqrt(double);
```

Now **sqrt(double)** can be called from a C or a C++ code fragment. The definition of **sqrt(double)** can also be compiled as a C function or as a C++ function.

Only one function of a given name in a scope can have C linkage (because C doesn't allow function overloading). A linkage specification does not affect type checking, so the C++ rules for function calls and argument checking still apply to a function declared **extern "C"**.

19.4 Bibliography（参考文献）

[Boost] *The Boost Libraries: free peer-reviewed portable C++ source libraries.* www.boost.org.

[C,1990] X3 Secretariat: *Standard – The C Language*. X3J11/90-013. ISO Standard ISO/IEC 9899-1990. Computer and Business Equipment Manufacturers Association. Washington, DC.

[C,1999] ISO/IEC 9899. *Standard – The C Language*. X3J11/90-013-1999.

[C,2011] ISO/IEC 9899. *Standard – The C Language*. X3J11/90-013-2011.

[C++,1998] ISO/IEC JTC1/SC22/WG21 (editor: Andrew Koenig): *International Standard – The C++ Language*. ISO/IEC 14882:1998.

[C++,2004] ISO/IEC JTC1/SC22/WG21 (editor: Lois Goldthwaite): *Technical Report on C++ Performance*. ISO/IEC TR 18015:2004(E) ISO/IEC 29124:2010.

[C++,2011] ISO/IEC JTC1/SC22/WG21 (editor: Pete Becker): *International Standard – The C++ Language*. ISO/IEC 14882:2011.

[C++,2014] ISO/IEC JTC1/SC22/WG21 (editor: Stefanus Du Toit): *International Standard – The C++ Language*. ISO/IEC 14882:2014.

[C++,2017] ISO/IEC JTC1/SC22/WG21 (editor: Richard Smith): *International Standard – The C++ Language*. ISO/IEC 14882:2017.

[C++,2020] ISO/IEC JTC1/SC22/WG21 (editor: Richard Smith): *International Standard – The C++ Language*. ISO/IEC 14882:2020.

[Cppcoro] *CppCoro – A coroutine library for C++.* github.com/lewissbaker/cppcoro.

[Cppreference] *Online source for C++ language and standard library facilities.* www.cppreference.com.

[Cox,2007] Russ Cox: *Regular Expression Matching Can Be Simple And Fast.* January 2007. swtch.com/˜rsc/regexp/regexp1.html.

[Dahl,1970] O-J. Dahl, B. Myrhaug, and K. Nygaard: *SIMULA Common Base Language.* Norwegian Computing Center S-22. Oslo, Norway. 1970.

[Dechev,2010] D. Dechev, P. Pirkelbauer, and B. Stroustrup: *Understanding and Effectively Preventing the ABA Problem in Descriptor-based Lock-free Designs.* 13th IEEE Computer Society ISORC 2010 Symposium. May 2010.

[DosReis,2006] Gabriel Dos Reis and Bjarne Stroustrup: *Specifying C++ Concepts.* POPL06. January 2006.

[Ellis,1989] Margaret A. Ellis and Bjarne Stroustrup: *The Annotated C++ Reference Manual.* Addison-Wesley. Reading, Massachusetts. 1990. ISBN 0-201-51459-1.

[Garcia,2015] J. Daniel Garcia and B. Stroustrup: *Improving performance and maintainability through refactoring in C++11.* Isocpp.org. August 2015.

[Friedl,1997] Jeffrey E. F. Friedl: *Mastering Regular Expressions.* O'Reilly Media. Sebastopol, California. 1997. ISBN 978-1565922570.

[Gregor,2006] Douglas Gregor et al.: *Concepts: Linguistic Support for Generic Programming in C++.* OOPSLA'06.

[Ichbiah,1979] Jean D. Ichbiah et al.: *Rationale for the Design of the ADA Programming Language.* SIGPLAN Notices. Vol. 14, No. 6. June 1979.

[Kazakova,2015] Anastasia Kazakova: *Infographic: C/C++ facts.*

[Kernighan,1978] Brian W. Kernighan and Dennis M. Ritchie: *The C Programming Language.* Prentice Hall. Englewood Cliffs, New Jersey. 1978.

[Kernighan,1988] Brian W. Kernighan and Dennis M. Ritchie: *The C Programming Language, Second Edition.* Prentice-Hall. Englewood Cliffs, New Jersey. 1988. ISBN 0-13-110362-8.

[Knuth,1968] Donald E. Knuth: *The Art of Computer Programming.* Addison-Wesley. Reading, Massachusetts. 1968.

[Koenig,1990] A. R. Koenig and B. Stroustrup: *Exception Handling for C++ (revised).* Proc USENIX C++ Conference. April 1990.

[Maddock,2009] John Maddock: *Boost.Regex.* www.boost.org. 2009. 2017.

[Orwell,1949] George Orwell: *1984.* Secker and Warburg. London. 1949.

[Paulson,1996] Larry C. Paulson: *ML for the Working Programmer.* Cambridge University Press. Cambridge. 1996. ISBN 978-0521565431.

[Richards,1980] Martin Richards and Colin Whitby-Strevens: *BCPL – The Language and Its Compiler.* Cambridge University Press. Cambridge. 1980. ISBN 0-521-21965-5.

[Stepanov,1994] Alexander Stepanov and Meng Lee: *The Standard Template Library*. HP Labs Technical Report HPL-94-34 (R. 1). 1994.

[Stepanov,2009] Alexander Stepanov and Paul McJones: *Elements of Programming*. Addison-Wesley. Boston, Massachusetts. 2009. ISBN 978-0-321-63537-2.

[Stroustrup,1979] Personal lab notes.

[Stroustrup,1982] B. Stroustrup: *Classes: An Abstract Data Type Facility for the C Language*. Sigplan Notices. January 1982. The first public description of "C with Classes."

[Stroustrup,1984] B. Stroustrup: *Operator Overloading in C++*. Proc. IFIP WG2.4 Conference on System Implementation Languages: Experience & Assessment. September 1984.

[Stroustrup,1985] B. Stroustrup: *An Extensible I/O Facility for C++*. Proc. Summer 1985 USENIX Conference.

[Stroustrup,1986] B. Stroustrup: *The C++ Programming Language*. Addison-Wesley. Reading, Massachusetts. 1986. ISBN 0-201-12078-X.

[Stroustrup,1987] B. Stroustrup: *Multiple Inheritance for C++*. Proc. EUUG Spring Conference. May 1987.

[Stroustrup,1987b] B. Stroustrup and J. Shopiro: *A Set of C Classes for Co-Routine Style Programming*. Proc. USENIX C++ Conference. Santa Fe, New Mexico. November 1987.

[Stroustrup,1988] B. Stroustrup: *Parameterized Types for C++*. Proc. USENIX C++ Conference, Denver, Colorado. 1988.

[Stroustrup,1991] B. Stroustrup: *The C++ Programming Language (Second Edition)*. Addison-Wesley. Reading, Massachusetts. 1991. ISBN 0-201-53992-6.

[Stroustrup,1993] B. Stroustrup: *A History of C++: 1979–1991*. Proc. ACM History of Programming Languages Conference (HOPL-2). ACM Sigplan Notices. Vol 28, No 3. 1993.

[Stroustrup,1994] B. Stroustrup: *The Design and Evolution of C++*. Addison-Wesley. Reading, Massachusetts. 1994. ISBN 0-201-54330-3.

[Stroustrup,1997] B. Stroustrup: *The C++ Programming Language, Third Edition*. Addison-Wesley. Reading, Massachusetts. 1997. ISBN 0-201-88954-4. Hardcover ("Special") Edition. 2000. ISBN 0-201-70073-5.

[Stroustrup,2002] B. Stroustrup: *C and C++: Siblings*, *C and C++: A Case for Compatibility*, and *C and C++: Case Studies in Compatibility*. The C/C++ Users Journal. July-September 2002. www.stroustrup.com/papers.html.

[Stroustrup,2007] B. Stroustrup: *Evolving a language in and for the real world: C++ 1991-2006*. ACM HOPL-III. June 2007.

[Stroustrup,2009] B. Stroustrup: *Programming – Principles and Practice Using C++*. Addison-Wesley. Boston, Massachusetts. 2009. ISBN 0-321-54372-6.

[Stroustrup,2010] B. Stroustrup: "New" Value Terminology.

[Stroustrup,2012a] B. Stroustrup and A. Sutton: *A Concept Design for the STL*. WG21 Technical Report N3351==12-0041. January 2012.

[Stroustrup,2012b] B. Stroustrup: *Software Development for Infrastructure*. Computer. January 2012. doi:10.1109/MC.2011.353.

[Stroustrup,2013] B. Stroustrup: *The C++ Programming Language (Fourth Edition)*. Addison-Wesley. Boston, Massachusetts. 2013. ISBN 0-321-56384-0.

[Stroustrup,2014] B. Stroustrup: C++ Applications.

[Stroustrup,2015] B. Stroustrup and H. Sutter: *C++ Core Guidelines*.

[Stroustrup,2015b] B. Stroustrup, H. Sutter, and G. Dos Reis: *A brief introduction to C++'s model for type- and resource-safety*. Isocpp.org. October 2015. Revised December 2015.

[Stroustrup,2017] B. Stroustrup: *Concepts: The Future of Generic Programming (or How to design good concepts and use them well)*. WG21 P0557R1.

[Stroustrup,2020] B. Stroustrup: *Thriving in a crowded and changing world: C++ 2006-2020*. ACM/SIGPLAN History of Programming Languages conference, HOPL-IV. June 2020.

[Stroustrup,2021] B. Stroustrup: *Type-and-resource safety in modern C++*. WG21 P2410R0. July 2021.

[Stroustrup,2021b] B. Stroustrup: *Minimal module support for the standard library*. P2412r0. July 2021.

[Sutton,2011] A. Sutton and B. Stroustrup: *Design of Concept Libraries for C++*. Proc. SLE 2011 (International Conference on Software Language Engineering). July 2011.

[WG21] ISO SC22/WG21 The C++ Programming Language Standards Committee: *Document Archive*. www.open-std.org/jtc1/sc22/wg21.

[Williams,2012] Anthony Williams: *C++ Concurrency in Action – Practical Multithreading*. Manning Publications Co. ISBN 978-1933988771.

[Wong,2020] Michael Wong, Howard Hinnant, Roger Orr, Bjarne Stroustrup, Daveed Vandevoorde: *Direction for ISO C++*. WG21 P2000R1. July 2020.

[Woodward,1974] P. M. Woodward and S. G. Bond: *Algol 68-R Users Guide*. Her Majesty's Stationery Office. London. 1974.

19.5 Advice（建议）

[1] The ISO C++ standard [C++,2020] defines C++.

[2] When choosing a style for a new project or when modernizing a code base, rely on the C++ Core Guidelines; §19.1.4.

[3] When learning C++, don't focus on language features in isolation; §19.2.1.

[4] Don't get stuck with decades-old language-feature sets and design techniques; §19.1.4.

[5] Before using a new feature in production code, try it out by writing small programs to test the standards conformance and performance of the implementations you plan to use.

[6] For learning C++, use the most up-to-date and complete implementation of Standard C++ that you can get access to.

[7] The common subset of C and C++ is not the best initial subset of C++ to learn; §19.3.2.1.

[8] Avoid casts; §19.3.2.1; [CG: ES.48].

[9] Prefer named casts, such as static_cast over C-style casts; §5.2.3; [CG: ES.49].

[10] When converting a C program to C++, rename variables that are C++ keywords; §19.3.2.

[11] For portability and type safety, if you must use C, write in the common subset of C and C++; §19.3.2.1; [CG: CPL.2].

[12] When converting a C program to C++, cast the result of malloc() to the proper type or change all uses of malloc() to uses of new; §19.3.2.2.

[13] When converting from malloc() and free() to new and delete, consider using vector, push_back(), and reserve() instead of realloc(); §19.3.2.1.

[14] In C++, there are no implicit conversions from ints to enumerations; use explicit type conversion where necessary.

[15] For each standard C header <X.h> that places names in the global namespace, the header <cX> places the names in namespace std.

[16] Use extern "C" when declaring C functions; §19.3.2.3.

[17] Prefer string over C-style strings (direct manipulation of zero-terminated arrays of char); [CG: SL.str.1].

[18] Prefer iostreams over stdio; [CG: SL.io.3].

[19] Prefer containers (e.g., vector) over built-in arrays.

A

Module std

（std 模块）

That is a big thing with an invention:
You have to have a whole system that works.
– J. Presper Eckert

- Introduction
- Use What Your Implementation Offers
- Use Headers
- Make Your Own **module std**
- Advice

A.1 Introduction（引言）

At the time of writing, **module std** [Stroustrup,2021b] is unfortunately not yet part of the standard. I have reasonable hope it will be part of C++23. This appendix offers some ideas of how to manage for now.

The idea of **module std** is to make all the components of the standard library simply and cheaply available by a single **import std;** statement. I have relied on that throughout the chapters. The headers are mentioned and named mostly because they are traditional and universally available, and partly because they reflect the (imperfect) historical organization of the standard library.

A few standard-library components dump names, such as **sqrt()** from **<cmath>**, into the global namespace. Module **std** doesn't do that, but when we need to get such global names we can **import std.compat**. The only really good reason for importing **std.compat** rather than **std** is to avoid messing with old code bases while still gaining some of the benefits from the increase in compile speeds from modules.

Note that modules, most deliberately, don't export macros. If you need macros, use **#include**.

Modules and header files coexist; that is, if you both **#include** and **import** an identical set of declarations, you will get a consistent program. This is essential for the evolution of large code bases from relying on header files to using modules.

A.2 Use What Your Implementation Offers（使用你的实现所提供的东西）

With a bit of luck, an implementation we want to use already has a **module std**. In that case, our first choice should be to use it. It may be labeled "experimental" and to use it may require a bit of setup or a few compiler options. So, first of all, explore if the implementation has a **module std** or equivalent. For example, currently (Spring 2022) Visual Studio offers a number of "experimental" modules, so using that implementation, we can define module **std** like this:

```
export module std;
export import std.regex;          // <regex>
export import std.filesystem;     // <filesystem>
export import std.memory;         // <memory>
export import std.threading;      // <atomic>, <condition_variable>, <future>, <mutex>,
                                  // <shared_mutex>, <thread>
export import std.core;           // all the rest
```

For that to work, we obviously have to use a C++20 compiler, and also set options to gain access to the experimental modules. Beware that everything "experimental" will change over time.

A.3 Use Headers（使用头文件）

If an implementation doesn't yet support modules or doesn't yet offer a **module std** or equivalent, we can fall back to using traditional headers. They are standard and universally available. The snag is that to get an example to work, we need to figure out which headers are needed and **#include** them. Chapter 9 can help here, and we can look up the name of a feature we want to use on [cppreference] to see which header it is part of. If that gets tedious, we can gather frequently used headers into a **std.h** header:

```
// std.h

#include <iostream>
#include<string>
#include<vector>
#include<list>
#include<memory>
#include<algorithms>
// ...
```

and then

```
#include "std.h"
```

The problem here is that **#include**ing so much can give very slow compiles [Stroustrup,2021b].

A.4 Make Your Own module std（制作你自己的 std 模块）

This is the least attractive alternative because it's likely to be the most work, but once someone has done it, it can be shared:

```
module;
#include <iostream>
#include<string>
#include<vector>
#include<list>
#include<memory>
#include<algorithms>
// ...

export module std;
export istream;
export ostream;
export iostream;
// ...
```

There is a shortcut:

```
export module std;

export import "iostream";
export import "string";
export import "vector";
export import "list";
export import "memory";
export import "algorithms";
// ...
```

The construct

```
import "iostream";
```

Importing a *header unit* is a half-way house between modules and header files. It takes a header file and makes it into something like a module, but it can also inject names into the global namespace (like **#include**) and it leaks macros.

This is not quite as slow to compile as **#include**s but not as fast as a properly constructed named module.

A.5 Advice（建议）

[1] Prefer modules provided by an implementation; §A.2.
[2] Use modules; §A.3.
[3] Prefer named modules over header units; §A.4.
[4] To use the macros and global names from the C standard, **import std.compat**; §A.1.
[5] Avoid macros; §A.1. [CG: ES.30] [CG: ES.31].